New Perspectives in Welding and Joining Processes of Metallic Materials

New Perspectives in Welding and Joining Processes of Metallic Materials

Guest Editors

Fabio Giudice
Cristina Scolaro

MDPI

Basel • Beijing • Wuhan • Barcelona • Belgrade • Novi Sad • Cluj • Manchester

Guest Editors
Fabio Giudice
Department of Civil
Engineering and Architecture
University of Catania
Catania
Italy

Cristina Scolaro
Department of Engineering
University of Messina
Messina
Italy

Editorial Office
MDPI AG
Grosspeteranlage 5
4052 Basel, Switzerland

This is a reprint of the Special Issue, published open access by the journal *Materials* (ISSN 1996-1944), freely accessible at: https://www.mdpi.com/journal/materials/special_issues/6HO2Q3U558.

For citation purposes, cite each article independently as indicated on the article page online and as indicated below:

Lastname, A.A.; Lastname, B.B. Article Title. *Journal Name* **Year**, *Volume Number*, Page Range.

ISBN 978-3-7258-7510-8 (Hbk)
ISBN 978-3-7258-7511-5 (PDF)
https://doi.org/10.3390/books978-3-7258-7511-5

Contents

About the Editors

Fabio Giudice

Fabio Giudice is Associate Professor of Metallurgy at the Department of Civil Engineering and Architecture of the University of Catania. Former Assistant Professor of Metallurgy at the same institution, he has been an Associate Researcher in Industrial Design, Mechanical Engineering and Metallurgy at the Department of Industrial Engineering of the University of Catania, and previously in Mechanical Design and Construction of Machines at the Department of Industrial and Mechanical Engineering of the same University. He graduated in Mechanical Engineering at the University of Catania, obtained a Master's Degree in Industrial Design at the Research Centre of the European Institute of Design in Milano, and a PhD in Structural Mechanics at the University of Catania. With research interests in the Mechanical Behavior and Microstructural Characterization of Metallic Materials, Optimal Materials Selection, Metallurgy of Processes based on Mobile Thermal Sources (Laser Welding, Metal Additive Manufacturing by Electron Beam Melting), Metallurgical Phenomena in the Manufacturing and Use of Components for Process Plant, Life Cycle Design and Design for X, and Product Design for the Environment, he has published 90 papers in these areas (in international and national journals, proceedings of international and national conferences, chapters on international books) and a book for a leading international publishing house.

Cristina Scolaro

Cristina Scolaro is Assistant Professor of Polymeric Materials for the Bioengineering course in the Bachelor's Degree in Industrial Engineering at the University of Messina. At the same university, she was a Postdoctoral Researcher in the PON-"Thalassa" Project, PO-FESR "Simare" Project, and "LIFE RESTART" Project. She has developed expertise in Metallurgy and Mechanical Metallurgy, completing specialized courses within the Industrial Engineering degree program at the University of Messina. She is a Professor of Mathematics and Science (Physics, Chemistry, Earth Sciences, Biology) and Mathematics and Physics for Secondary Schools in Messina. She graduated in Physics from the University of Messina, where she earned a PhD in Physics and a PhD in Engineering and Chemistry of Materials and Constructions. She completed a two-year degree in Psychopedagogical Methodologies for the Management of Teaching and Learning in the Educational Field, in the scientific-disciplinary area of secondary school at the University for Foreigners "Dante Alighieri" in Reggio Calabria. Her research interests include Materials Science and Technology, particularly polymeric and metallic materials, and their characterization for advanced engineering applications, such as structural, mechanical, and thermal analysis. She focuses on surface (wettability and roughness) and bulk (static, rheological, and calorimetric) techniques, essential for the optimization of biomaterials and components in the biomedical, industrial, and energy sectors. She has published 60 articles in these fields (in international and national journals, and in international and national conference proceedings).

Preface

In all fields of engineering applications, welding and joining techniques for metallic components play a key role in ensuring the required performance, quality and safety in service. They have a direct influence on the primary properties of equipment and structures, which, more and more frequently, should satisfy an increasingly broad set of requirements and multiple functions under operating conditions.

The continuous development of new metal alloys promotes research on advanced welding and joining technologies. Given the vast and diverse range of requirements and functions to be satisfied, investigations into the compatibility and weldability of materials, and also into the metallurgical effects of joining processes and parameters on their final microstructure and properties, are an essential phase in selecting and setting the most efficient solutions.

In this regard, research to thoroughly understand the process mechanisms and related metallurgical phenomena on a scientific basis is required to be continuously and intensively carried out, and further work is still required in this field to interpret the correlation between process parameters, material microstructure, and joint efficiency. In addition, the current strong push towards environmental protection policies also raises the question of how welding and joining technologies can fit into the more general framework of the environmental sustainability of manufacturing processes.

This Special Issue presents research articles and reviews from academia, research institutions, and industry with the aim of providing an overview on recent advances in welding and joining processes of metallic materials and outlining the current perspectives in the field.

Fabio Giudice and Cristina Scolaro
Guest Editors

materials

Review

A Review on Metallurgical Issues in the Production and Welding Processes of Clad Steels

Fabio Giudice [1,*], Severino Missori [2], Cristina Scolaro [3] and Andrea Sili [3,*]

1 Department of Civil Engineering and Architecture, University of Catania, 95123 Catania, Italy
2 Department of Industrial Engineering, University of Rome-Tor Vergata, 00133 Roma, Italy; missori@uniroma2.it
3 Department of Engineering, University of Messina, 98166 Messina, Italy; cscolaro@unime.it
* Correspondence: fabio.giudice@unict.it (F.G.); asili@unime.it (A.S.)

Abstract: Carbon and low-alloy steel plates clad with stainless steel or other metals are a good choice to meet the demand for cost-effective materials to be used in many corrosive environments. Numerous technical solutions are developed for the production of clad steel plates, as well as for their joining by fusion welding. For thick plates, a careful strategy is required in carrying out the multiple passes and in choosing the most suitable filler metals, having to take into account the composition of the base metal and the cladding layer. The specificity of the different processes and materials involved requires an adequate approach in the study of the metallurgical characteristics of clad steel, thus arousing the interest of researchers. Focusing mainly on ferritic steel plates clad with austenitic steel, this article aims to review the scientific literature of recent years which deals with both the production and the fusion welding processes. The metallurgical issues concerning the interfaces and the effects of microstructural characteristics on mechanical behaviour and corrosion resistance will be addressed; in particular, the effects on the fusion and thermally affected zones that form during the fusion welding and weld overlay processes will be analysed and discussed.

Keywords: clad steel; filler metal; arc welding; laser beam; welding passes; weld zone; solidification mode; heat-affected zone

1. Introduction

Cladding of carbon and low-alloy steel plates with a layer of stainless steel, aluminium or titanium alloy is a cost-effective solution to the growing demand for quality materials coming from various industrial fields, such as the petrochemical, energy production or shipbuilding fields. In this way, the mechanical properties expected from the backing steel can be combined with the corrosion resistance of the cladding metals. In recent decades, to achieve good performance even in aggressive environments, ferritic steels clad with austenitic steels have been successfully utilised in the petrochemical and energy industries for the production of vessels and heat exchangers and for making longitudinally welded clad pipes.

The cladding material can be applied through surface welding (by single- or multi-pass conventional arc welding processes), plate remelting by laser beam welding (LBW), explosion bonding and solid-state welding by coextrusion of cylinders or hot rolling bonding of plates. The last method is suitable for large-scale industrial production (for a review, see the article by Wang et al. [1]). In particular, the hot rolling process is widely utilised to produce cladding layers of austenitic steel on carbon steel plates, which is a cheap solution to meet the requirements of corrosion resistance typical of the heat exchangers and pressure vessels. These plates combine the good mechanical properties of low-cost carbon steel with the high corrosion resistance and heat resistance typical of stainless steel; furthermore, the strong interface bonding makes clad plates fit for further forming processes by plastic deformation or welding [2].

Even if austenitic stainless steel clad plates are the most commonly used for structures in corrosive environments, cladding metals can also be selected from a wide variety of alloys, which provide specific properties expanding the fields of use of the base material. In particular, joining aluminium to steel is of high economic and technical interest for many industries, such as shipbuilding, since it combines the light weight of Al with the low cost and the high structural strength of steel. Transition joints between structural steel and aluminium alloys are currently produced by explosive welding, which is a joining process where the high strains and temperatures, acting in a short period of time, make it particularly suitable for dissimilar materials with very different metallurgical properties [3].

Generally, clad steels show interfaces with strong metallurgical bonding, even if some critical issues have been highlighted in the literature: for example, regarding stainless steel clad plates produced by hot rolling, the article by Dhib et al. [4] deals with the formation of a hard carburized layer due to the carbon diffusion towards the austenitic stainless steel, in which the precipitation of Cr-carbides occurs, with harmful consequences for corrosion resistance [5].

The fundamental requirement for clad steel weldments is to obtain a continuous layer that maintains the corrosion-resistant characteristics of the cladding material; however, in fusion welding, the significant differences in composition between the deposited layers are a source of drawbacks that could make the use of clad steels challenging. As documented in the literature (see the review by Wang et al. [1] and, more specifically, the articles cited in Section 3), several procedures based on the use of filler wire, applied via conventional arc welding or hybrid laser beam/arc welding, have been successfully developed over the years. However, in some cases they have been affected by issues that are essentially due to dilution phenomena between base steel and cladding alloys, as well as between adjacent layers obtained with sequential passes [6]. In thick plates, arc welding is carried out with multiple passes, which involve dilution phenomena and heating cycles, affecting the weld composition and in general the metallurgical characteristics of the fusion zone (FZ) and heat-affected zone (HAZ). In this regard, the addition of consumable inserts, interposed between the edges of two butt-positioned clad plates, has been proposed as an advantageous solution for LBW in a single pass to obtain narrow welds with deep penetration [7]. The purpose of this article is to outline a comprehensive analysis of the most recent scientific literature on issues relating to clad steel, focusing mainly on ferritic steel plates clad with austenitic steel.

The metallurgical features of both as-produced plates and welds will be specifically highlighted for their relevance in terms of mechanical properties and corrosion resistance. Therefore, the production processes as well as the various fusion welding techniques will be reviewed in Sections 2 and 3, respectively, to analyse the effects on clad steel properties. The main metallurgical issues in weld overlay and fusion welding will be debated in Section 4, with particular regard to those arising from dilution and solidification modes. Finally, conclusions and future research directions will be outlined in Section 5.

2. Metallurgy of As-Produced Clad Steel Plates

2.1. Hot Rolled Clad Steel Plates

Hot rolling is widely used for manufacturing clad plates, as it is an economic and efficient process for mass production. The thickness of the cladding layer, in general austenitic stainless steel, can be up to 6–7 mm, while that of the base carbon or low-alloyed steel is usually three or four times thicker [1]. With reference to Figure 1, rolling is performed on a pack of plates arranged in a sandwich-type composition: two sets of cladding layers/base steel are symmetrically assembled with an intermediate separating layer, then four bars are welded all around their edges to form a sealed chamber in which a vacuum can be obtained through an exhaust hole connected to a vacuum pump [8]. In this way, the formation of oxides, with consequent deterioration of the interface bonding, is avoided. An interlayer between the base steel and cladding is also added to prevent the formation of intermetallic compounds, thereby improving the bonding strength of the composite plate [9].

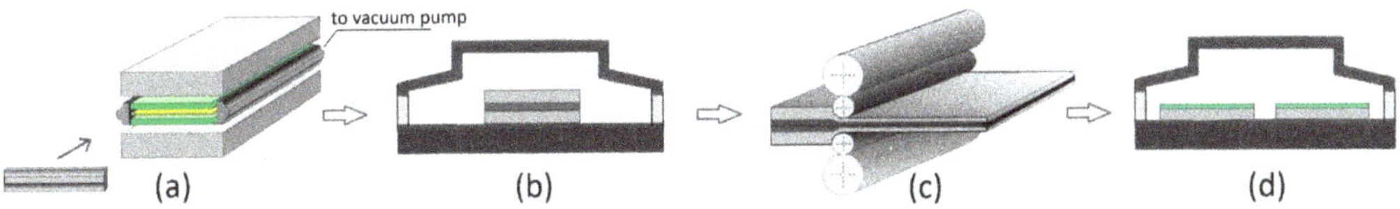

Figure 1. Hot rolling process of two sets of base steel (grey) and cladding layer (green) separated by an interposed layer (yellow): (**a**) assembling of the "sandwich"-type composition; (**b**) preheating; (**c**) hot rolling; (**d**) final annealing of the clad plates.

During the bonding process, the plates undergo some thermal cycles, which in the case of ASTM A283 grade C carbon steel (base metal) and AISI 316 austenitic steel (clad layer) consists of the following steps [10]:

- Preheating of the assembled pack, with a slow increase from room temperature up to 1230 °C, maintained for 5 h;
- Hot rolling between 1230 and 850 °C, followed by air cooling;
- Disassembling of the rolled pack;
- Final annealing in the range 920–950 °C, followed by air cooling.

Hot rolling gives rise to a continuous bound between the two metals, which maintain their composition unchanged, except in a narrow band at the interface where the permanence at high temperature activates diffusion. A thin and slight wave-like interface, the so-called cladding line, separates the ferritic/pearlitic microstructure of the base carbon steel from the typical austenitic microstructure of the cladding layer. The non-perfect linearity of the interface is a typical phenomenon of instability, due to the high rolling temperature and reduction ratio [2].

An ideal cladding interface should be integral and continuous without any defects, such as inclusions or unbonded areas which could affect the mechanical properties, as shown by Li et al. [11], who studied the interfacial fracture evolution in austenitic stainless steel clad plates.

The interfacial bonding strength is affected by several factors, such as the bonding temperature, the deformation reduction ratio, the roll speed and the interfacial oxidization. In this regard, Zhu et al. [12] found that the interface bonding strength and interface toughness can be affected by the presence of oxides positioned along the cladding line, which may act as alternative crack propagation paths in shear tests. Complete metallurgical bonding can be obtained at high temperatures and in a protective atmosphere or in a vacuum.

Compared with cold roll bonding, hot rolling allows easy bonding of the two metals with a slight deformation in thickness, due to the effects of plastic deformation and recrystallization. However, the bond strength decreases drastically when the metal surfaces are oxidized; therefore, their preparation by appropriate mechanical polishing and chemical etching [10] is a key factor for the optimal success of the bonding process [13].

The permanence at a high temperature of the plates generates diffusion phenomena across the cladding line. The substitutional alloy elements, of which the austenitic layer is rich, diffuse towards the backing carbon steel, while the small interstitial atoms of carbon migrate more quickly in the opposite direction, thus affecting a larger zone [4]. As shown by the diffusion profile in Figure 2, the interface between the two steels is characterised by a diffusion layer, 20 μm thick, where the contents of Cr, Ni and Mn decrease linearly from the stainless steel side to the carbon steel one, whereas the content of Fe has an opposite trend; furthermore, the mobile carbon atoms diffuse over a longer distance towards the austenitic steel [5].

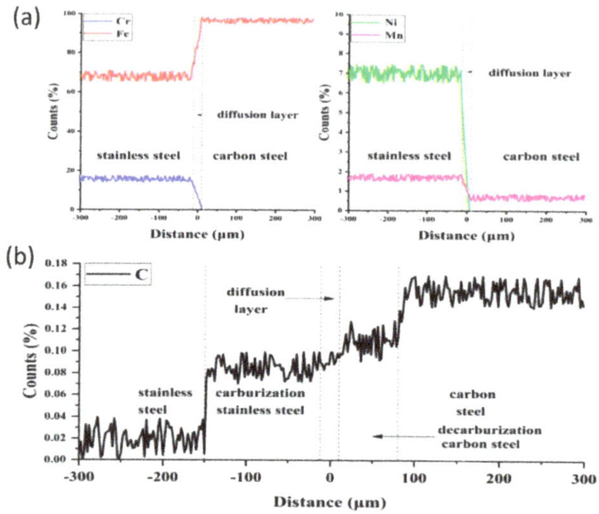

Figure 2. Results of electron probe micro-analyser measurements across the interface SUS304/Q235: (**a**) Cr, Fe, Ni and Mn; (**b**) carbon. Reproduced from [5].

Due to diffusion in hot rolled clad plates, a decarburized zone consisting of coarse ferritic grains (in which pearlite is absent) and a carburized layer are formed on the carbon steel and austenitic stainless steel sides, respectively. With reference to Figure 3, the following areas can be identified at the carbon steel/austenitic steel interface [5]: a decarburized layer near the interface, mainly composed of ferrite (A zone); carbon steel far from the interface consisting of ferrite and pearlite aligned along the rolling band (B zone); a carburized layer at the austenitic steel side, characterised by Cr-carbide precipitation at the grain boundary (C zone); and an unaffected austenitic steel microstructure (D zone).

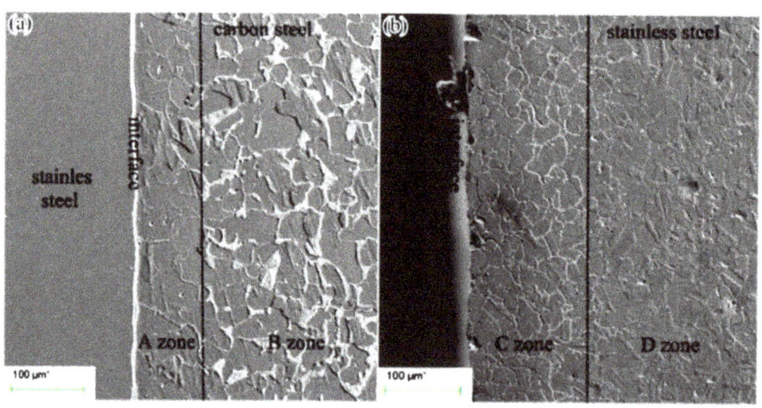

Figure 3. Clad plate interface: (**a**) Q235 carbon steel side; (**b**) SUS304 austenitic steel side. Reproduced from [5].

The diffusion coefficient of carbon is several orders of magnitude larger than that of chromium; therefore, the carbon atoms readily diffuse through the bonding interface, whereas the diffusion of Cr is much more difficult [14]. Hence, carbon atoms diffuse into the austenitic stainless steel layer, where Cr-carbide formation occurs during cooling in the sensitizing temperature range 550–850 °C, causing a sharp increase in hardness near the cladding line, and maintaining high values within the carburized layer at the austenitic steel side. This is confirmed by the results of the Vickers microhardness surveys performed along a line transverse to the interface, such as in the case shown in Figure 4: the hardness peak increases up to 1475 K, as long as the effect due to carbon diffusion prevails, whereas higher temperatures cause a softening of the matrix and consequently a reduction in hardening, which becomes evident at 1575 K [15].

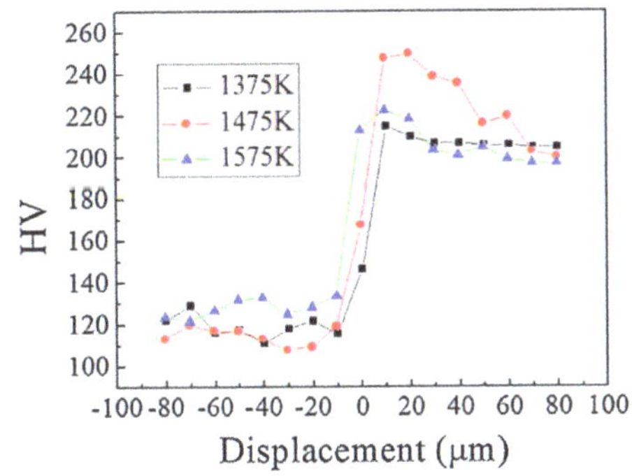

Figure 4. Vickers microhardness profile along a transversal line across the interface, from the Q235 carbon steel side (on the left) to the SUS 304 austenitic steel side (on the right), for three different rolling temperatures. Reproduced from [15] with permission from Elsevier.

In any case, the tensile properties of the hot rolled plates change significantly as a function of the rolling temperature or time. To overcome this issue, Yang et al. [16] proposed a new method based on a liquid–solid bonding process, in which stainless steel is deposited through a melting crucible onto a carbon steel plate. The clad plate, subsequently hot rolled at 900 °C, results in a superior combination of high shear strength and ductility of the cladding interface.

Cr-carbide precipitation causes Cr depletion at the grain boundary. Here, the susceptibility to intergranular corrosion can significantly increase, especially when the Cr content is reduced to values lower than 12 wt.% [17]. The effects of sensitization to intergranular corrosion on the austenitic steel side near the cladding line are shown by the micrograph in Figure 5 [18], taken after the ASTM Test A262-Practice E [19] was carried out.

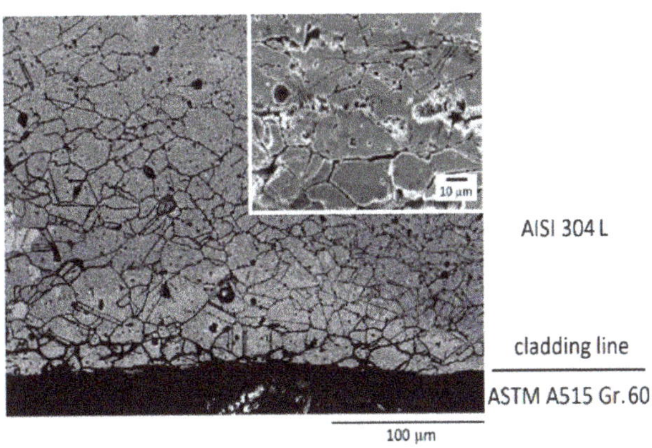

Figure 5. Sensitization to the intergranular corrosion of the carburized zone near the interface of carbon steel/austenitic steel (the higher magnification in the box shows the effect of corrosion immediately close to the cladding line). Reproduced from [18].

The width of the zones affected by diffusion depends on the processing temperature, as documented by Liu et al. [20] in the case of Q235 carbon steel plates with SUS 304 austenitic steel cladding layers, assembled in a vacuum, maintained for 2–3 h at three different temperatures (1100, 1200 and 1300 °C) and then hot rolled. They observed a reduction in ultimate tensile strength (σ_{UTS}) and an increase in elongation (ε) as the preheating temperature increased from 1100 to 1300 °C: in particular, σ_{UTS} varied from 578 $\pm$ 8 to 528 $\pm$ 8 MPa and ε from 46 $\pm$ 3 to 60 $\pm$ 3%. This is due to the formation of a strong interface that delays the propagation of delamination cracks and localized necking.

Considering that the hot rolled plates could suffer deformation-related defects, vacuum diffusion bonding is a good alternative for small productions. This process is performed in a vacuum chamber by applying pressure at a high temperature. Therefore, it allows one to manage the main working parameters, such as temperature, pressure and time, obtaining further improvements in the interfacial bonding strength, as demonstrated by Li et al. [21]. The authors achieved good metallurgically bonded interfaces in stainless steel clad plates; they experimented with the effects of the bonding temperature (ranging from 700 to 1100 °C) and time (from 1 to 4 h), demonstrating that the best mechanical properties were obtained when the manufacturing process was carried out under a pressure of 15 MPa, at 800 °C and for 120 min.

Concerning post-processing heat treatment, it is noteworthy to consider that stainless steel generally requires temperatures above 1000 °C to dissolve deleterious secondary phases which may form in the range 600–100 °C, while the mechanical properties of carbon steel get worse when treated above 1000 °C because of grain growth. Therefore, it is necessary to determine, for each specific case, the optimal treatment conditions for achieving a balance between the properties of the two metals. For example, Song et al. [22] developed a heat treatment for a clad steel to be utilised for hull structures in shipbuilding (S32750 cladding metal/EH40 base metal EH40), consisting of water quenching after treatment at 1080 °C for 1 h/in and tempering at 550 °C for 1–2 h followed by air cooling.

2.2. Explosion-Welded Clad Steel Plates

It is a fact that joining aluminium to steel is challenging for metallurgy, due to the wide difference in their material properties (especially in the melting point) and the high susceptibility to generating brittle intermetallic phases, which make the use of conventional welding techniques very difficult.

For these reasons, both the volume of the interacting materials and the permanence at a high temperature should be minimised. High-velocity impact processes meet these requirements; moreover, they can be performed without the addition of an external heat input (for a review, see the article by Wang et al. [23]).

Among these processes, explosive welding, in which a pressure wave pushes with high velocity the so-called flyer plate against the back or base plate, is currently utilised to produce joints between metals that are otherwise non-weldable. The exceptional mechanical properties of the weld depend on the way the two surfaces are brought into contact, as the exchange of valence electrons makes the formation of interatomic bonds possible. In any case, explosive welding is characterised by an extremely short duration, which minimises heat dissipation and prevents the formation of the heat-affected zone (HAZ) typical of fusion welding [24].

Explosive welding of plates is generally carried out under inclined or parallel mode, as shown in Figure 6; a buffer was used to prevent the flyer plate from being damaged by the explosion. The inclined setup, under the initial angle α, was introduced first, whereas the parallel one was developed later to weld large plates with a pre-determined standoff distance. The jet sweeps away the oxide films on the surfaces of the two metallic plate metals, favouring the formation of metallurgical bonds on an atomic distance scale [25].

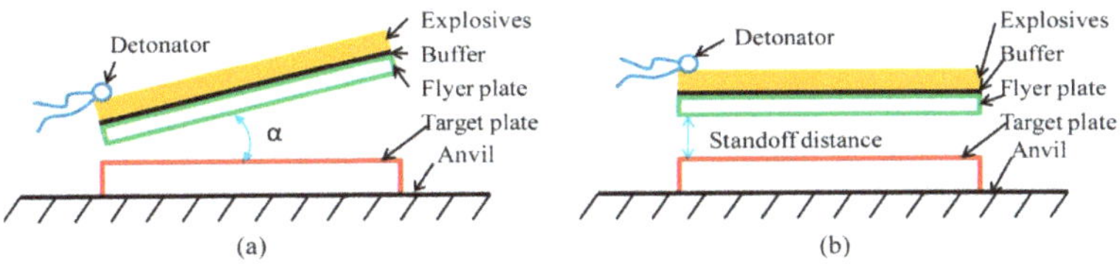

Figure 6. Sketch of the experimental setups for explosive welding: (**a**) inclined; (**b**) parallel. Reproduced from [23].

The transition joints between structural steel and Al alloy are useful in shipbuilding and in general when it is necessary to connect structures made of such different metals [26]. For their production, the best results are achieved using an explosive mixture (with low detonation velocity and a low explosive ratio) and an intermediate layer made of commercial pure aluminium [27].

This hybrid connection is becoming increasingly successful in shipbuilding as it allows the joining in turn, by conventional fusion welding, of the steel substrate to the steel hull and similarly the Al alloy layer to any light alloy superstructures (Figure 7).

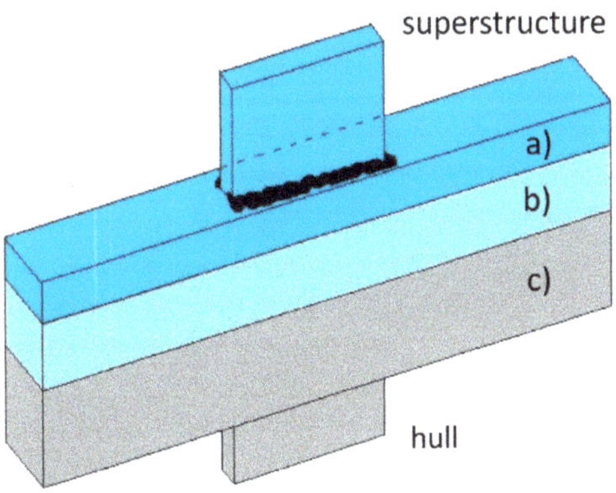

Figure 7. Triclad joint: (**a**) AA5083 Al alloy; (**b**) AA1050 Al interlayer; (**c**) ASTM A516 Gr 55 structural steel.

Several articles in the literature deal with the thermomechanical conditions and the process parameters: in the case of the Al-Fe system, Carvalho et al. [28] showed how the detonation velocity (experimentally measured in the range 2000–3000 m/s), with the calculated values of the impact velocity (about 300–400 m/s) and collision angle (from 7 to 10°), influences the final microstructure with the formation of embrittling intermetallic phases.

The morphology of the weld interface has been characterised by many authors. In some cases, the pressure shockwave due to the explosion causes a wavy profile, which has been documented in the literature for different joints, for example Al/Al alloy (Figure 8a) [27], CuCrZr/316L [29], AISI 316/Q235 B [30] and Cu-DHP/AISI 304L [31], whereas, in other cases, such as for an Al/carbon steel joint [27], the interface is almost flat (Figure 8b). For the conditions that lead to the formation of a wavy or straight interface, see also the numerical approach carried out by Ayele et al. [32] for a bimetallic composition (Al/carbon steel), and by Campanella et al. [3] for a joint with three dissimilar materials (AA5083 aluminium alloy and A516 steel, with an intermediate layer made of AA1050 aluminium alloy).

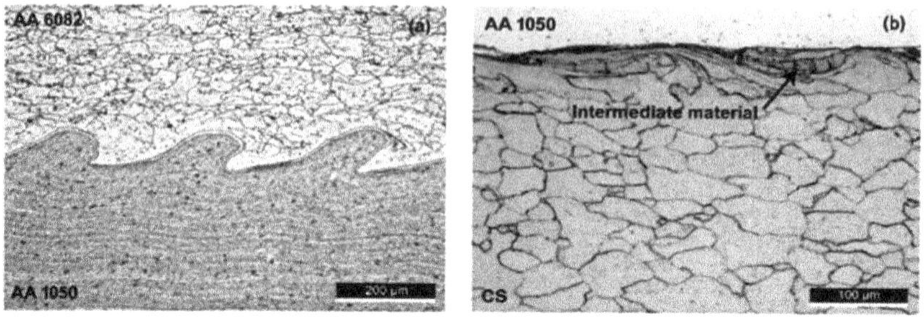

Figure 8. Micrographs of the weld interface: (**a**) AA6082 AA1050 interface; (**b**) AA1050/carbon steel interface. Reproduced from [27].

In this regard, Carvalho et al. [33] demonstrated that the AA 1050 interlayer improves the weldability of the AA 6082 flayer plate to the AISI 304 base plate, while it has no beneficial effects on the mechanical properties of the aluminium-to-carbon steel welds. However, the formation of embrittling intermetallic compounds can be observed at the interface between aluminium and steel, as indicated in Figure 7b. Therefore, attention must be paid, as these embrittling phases could grow in the case of subsequent thermal cycles resulting from any welding processes [34].

Explosive welding has proven to be suitable for joining aluminium and steel plates on large scales up to metres. It is currently utilised in shipbuilding to produce the Triclad transition joint consisting of ASTM A516 Grade 55 structural steel as the base plate and AA 5086 or AA 5083 aluminium alloy as the flyer plate, with AA 1050, almost pure aluminium, as the interlayer [35].

In the Triclad joint, as shown in [36], the interface on the side AA1050/A516 Grade 55 shows a linear profile with some particles recognized as an Al-Fe intermetallic phase, while on the opposite side, it is characterised by a wavy morphology that interlocks the Al/Al alloy interface, resulting in an increase in toughness and strength.

The wavy morphology depends on the process conditions, as demonstrated by Campanella et al. [3], who developed a numerical method for determining the parameters that affect the peak height and wavelength in the Triclad joint, as well as by Lee et al. [37], who simulated the wave formation in the Cu-Ti system.

2.3. Weld Overlaid Clad Steel Plates

For several decades, the weld surface cladding process of steel tubes or plates has been performed mainly by depositing the filler metal in the form of wire, usually molten by conventional arc welding [38]. This process, also called weld overlay, provides a cost-effective application of a layer at least 3 mm thick, in agreement with [39], whose final composition is determined by dilution between the filler and base metal. The overlaid layer is characterised by strong metallurgical bonding with the base material, due to the interpenetration of the deposited and supporting materials, which can be obtained with various traditional processes [40]; however, the formation of partially diluted zones (PDZs) at the interface between the cladding layer and base steel could be a potential issue, which will be discussed in Section 4.

The quality of cladding is affected by the weld bead geometry, which in turn depends on process parameters of arc welding, such as current, voltage and traveling speed, as highlighted by Saha et al. [41]. They performed the deposition of AISI 316 weld beads on a low-alloy structural steel plate using the gas metal arc welding (GMAW) process with 100% CO_2 as a shielding gas, experimenting with different combinations of current, voltage and arc travel speed, chosen so that the heat input increases from 0.35 to 0.75 kJ/mm. The GMAW process was also utilised by Aslam et al. [42] to deposit an AISI 304 layer on the surface of a mild steel plate. These authors developed a transient thermal numerical model to identify the effect of the welding parameters on the clad bead geometry.

Many authors have carried out experimental works to characterise the cladding obtained by weld overlay (see [43] for an overview of the different processes utilised and the layer characterisation): e.g., Sowrirajan et al. [44] investigated the effects of weld dilution on the thermal conductivity of austenitic stainless steel layers deposited by flux-cored arc welding (FCAW) on structural steel plates; Moreno et al. [45] focused their experimental work on the effects of FCAW parameters to qualify a martensitic weld bead deposited on AISI 1020 base steel; and Cattivelli et al. [46] studied any potential connection between the residual stresses generated during cladding through submerged arc welding (SAW), those generated during the post-weld heat treatment (PWHT) and the propensity for underclad cracking.

In general, the composition of a weld overlaid layer is characterised by a certain amount of dilution, due to the contribution of both the filler and base metal. Recently, Mattias et al. [47] developed a process of gas tungsten arc welding (GMAW) for depositing through three passes, on an ASTM A516 Gr. 70 carbon steel plate, a layer of super austenitic stainless steel containing 6% of Mo, the result of which was able to provide acceptable corrosion performance. The authors obtained a layer with a global dilution of 11% through the application of a low heat input, achieving high cooling rate and helping to minimise the secondary phases in agreement with the results shown in [48] on the solidification and segregation characteristics of a super austenitic stainless steel. The formation of cracks, due to metal shrinkage, could occur during the cooling stage. This drawback can be avoided by adding a buffer layer along with the multilayer deposition or by performing appropriate preheating of the electrode [43].

The electroslag welding (ESW) process is proposed as a valuable alternative for overlay welding applications, due to the possibility of obtaining a thinner cladding resulting in high productivity and significant cost reduction. This process was used in [49] to deposit a single layer of AISI 904 L on a substrate of ASTM A516 Gr. 70. The authors demonstrated that, due to its superior deposition rate, the ESW process can provide weld claddings with adequate microstructure and corrosion resistance together with increased productivity and cost reduction compared to the usual arc welding processes.

Recently, the gas metal arc welding process with rotating electrode (GMAW-RE) has shown to be promising, since the rotational movement of the wire-electrode provides more homogeneous weld bead profiles, suitable to weld overlaying, as shown by Costa et al. [50]. In their work, the feasibility of the GMAW-RE in depositing a single layer of Inconel 625 weld cladding with suitable properties was investigated. They obtained a homogeneous weld penetration profile, low dilution ratio and adequate reinforcement, thanks to an austenitic microstructure containing a low number of secondary phases.

The hot wire gas tungsten arc welding (GTAW-HW) process benefits from preheating the filler wire before reaching the melt pool. This process has proven to provide high quality, versatility and low cost with an appropriate setting of process parameters [51]. It was applied by Conzaga et al. [52] to weld overlay an API 5L X65 steel pipe with a 70%Ni30%Cu alloy by HW-GTAW. The microstructural and mechanical characterisation performed by the authors demonstrated promising results for applications in seawater systems.

In the usual weld overlay processes, the filler is deposited on the base metal by melting a consumable wire via an electric arc or electro slag process. Nowadays, laser direct metal deposition is proving to be a prominent technique, in which the filler, in the form of a metallic powder, is sprayed over the substrate and sintered (Figure 9a). A laser beam is a flexible tool for melting metals in different forms; therefore, a possible filler for cladding metal could be waste materials in the form of a plate, machined at a given thickness. In this regard, Bunaziv et al. [53] carried out the remelting of 2.0 mm thick 316 L stainless steel plates placed on a carbon steel substrate, using a high-power fibre laser beam (Figure 9b). This process, which allows one to achieve an acceptable corrosion resistance, is inexpensive and promising since scrap metals can be reused as filler material.

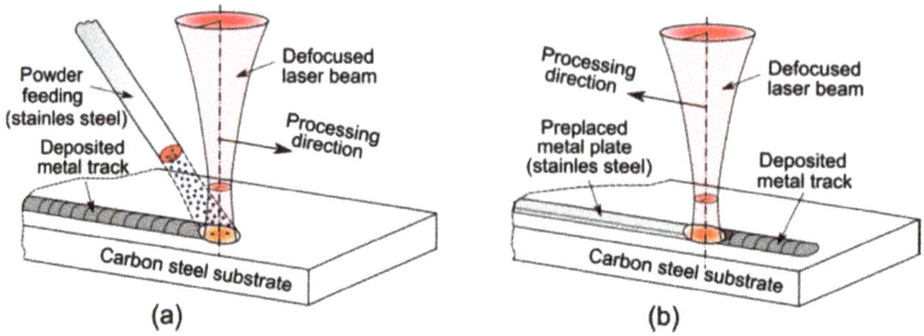

Figure 9. Weld overlay of carbon steel plates: (**a**) conventional processing using wire or powder as feedstock; (**b**) modified processing using a scrap plate as feedstock. Reproduced from [53].

As shown in Figure 10, in laser beam overlay, the focal point position (FPP) is a crucial parameter to achieve the optimal track geometry, avoiding the presence of melted zones between two passes and minimising dilution with the carbon steel substrate.

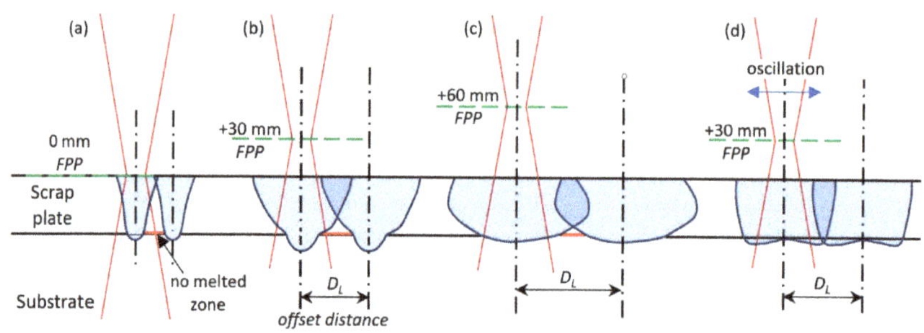

Figure 10. Geometries of two adjacent tracks for different values of FPP: (**a**) narrow track and insufficient overlapping; (**b**) optimization of tracks by defocused laser beam; (**c**) highly defocused laser beam; (**d**) optimal tracks obtained with laser beam oscillations. Reproduced from [53].

In recent years, due to its excellent stability, precise heat input and high power density, laser beam welding has also proven to be a promising technique for underwater processes, such as welding, weld cladding or laser remelting to repair and improve metal surfaces of marine structures, as shown in the experimental work of Li et al. [54].

When working underwater, the irradiated water can strongly shield the laser beam, resulting in a reduction in the absorption efficiency of the laser upon the workpiece surface. Under this condition, the formation of a dry region that acts as a channel around the laser beam is necessary for a successful process. Therefore, a drainage device is needed to produce a dry area on the surface of the damaged marine structure to keep water away and ensure operational flexibility and high manufacturing quality. Wang et al. [55] designed an in-house gas curtain nozzle, which was coupled with the laser cladding head as one underwater production tool. It was utilised to repair a pre-damaged surface of a mill-annealed HSLA-100 steel plate by weld overlaying the filler material consisting of HSLA-100 powder in the form of gas-atomized spherical particles. Other different technical solutions have been developed in the literature, such as drainage methods based on a curtain of gas or water convoyed onto the workpiece surface (for a review on this topic, see [56]).

3. Fusion Welding of Clad Steel

3.1. Arc Welding Processes (Hybrid Multi-Passes)

Clad steel plates are usually joined by fusion welding, because other types of fasteners would require unacceptable machining, such as drilling holes for bolting, which could leave the base metal exposed to environmental etching. Currently, various welding methods, such as GTAW, GMAW, SAW or shielded metal arc welding (SMAW), are commonly carried out by depositing multiple layers using traditional multi-pass procedures, as reported below.

In any case, fusion welding of stainless steel clad plates is challenging due to the great diversity in chemical compositions, microstructures and mechanical and physical properties between the substrate and the cladding metal. For this reason, clad plates are rarely welded with a single procedure, even if with different filler metals, such as in [10] where SMAW was used for both the base metal and cladding layer.

In multi-pass processes, specific working conditions should be applied at the base steel and cladding alloy level. In most cases, welding is performed by different methods, so as to use the one that is most appropriate for the substrate and cladding layer, respectively [57,58].

As a matter of fact, there are several issues due to the presence of two metals, each one with its peculiar characteristics, and specific requirements must be considered in the choice of the process setup. Firstly, the effects due to the different values of the linear expansion coefficient (about 12×10^{-6} and 17×10^{-6} °C^{-1} for carbon steel and austenitic steel, respectively) and thermal conductivity (about 45 and 16 W/(°C·m) for carbon steel and austenitic steel, respectively) could lead to the onset of residual stress and distortion, as documented in [59].

Furthermore, inhomogeneity in the weld composition could occur since the deposited layers undergo dilution with each other, as well as with the cladding alloy and base steel; in addition to this, the formation of diffusion zones of carbon and other alloy elements leads to harmful consequences for the mechanical and corrosion behaviour of the weld. To address these issues, various welding methods have been proposed; they are characterised by the combination of two or more welding processes and filler materials depending on the involved metals, as documented in some significant articles from recent years, cited below. These are essentially composite procedures, defined with the term "hybrid", being based on multiple passes with different arc welding processes and consequently with the deposition of multiple layers. In general, the welding sequence starting from the base steel provides better joint performance [6].

Dhib et al. [10] welded together two A283/A316 clad plates by a multi-pass and multilayer SMAW process, using three filler metals: A283 for the layers close to the A283 base steel, 316 for the layers close to the 316 cladding steel and 309L for the intermediate layers. They compared three different sequences, each one with a number of passes ranging from 7 to 15, obtaining the best results in terms of toughness when performing the 1st passes to seal the base steel. An et al. [60] showed that stainless steel clad plates can be successfully welded using GMAW (with ER70S-6 wire) and GTAW (with ER309L wire) for the Q235 base steel and the AISI 304 clad steel, respectively. They adopted a hybrid procedure in three passes: first, carbon steel backing welding, followed by carbon steel covering welding and, finally, stainless steel covering welding (Figure 11).

In [61], a procedure for welding two plates of Q235 carbon steel clad with AISI 304 austenitic steel (total thickness 7.2 mm) was used experimentally, depositing a transition layer of a suitable filler material to limit Cr and Ni diffusion and the consequent formation of a brittle martensitic zone. First, two passes of GMAW (ER50-6 filler) were carried out on the carbon steel side, then a transition layer was deposited using GTAW with E309L as a filler, or SMAW with the same filler. Finally, the last pass of SMAW (E308 filler) was performed to complete the clad steel side.

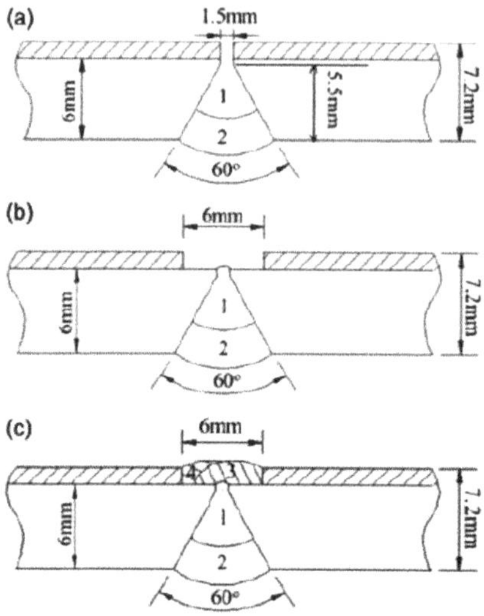

Figure 11. Welding sequences for two butt-positioned plates (Q235 base steel and AISI 304 cladding layer): (**a**) carbon steel backing welding and carbon steel covering welding; (**b**) cleaning the stainless steel cladding; (**c**) stainless steel covering welding. Reproduced from [60] with permission from Taylor & Francis Ltd.

In [62], two passes of GTAW were performed on the AISI 321 cladding layer, first with E347 filler for the root pass and then with ER309 filler for the transition layer. Finally, the Q345R carbon steel side was completed using four passes of the SMAW process with E309-16 filler, or otherwise with E4315 to save costs.

In the works by Ghorbel et al. [57,58], three passes of SMAW (E7018 filler) were performed on the A283 Gr C base low-carbon steel; then, the weld was completed by means of the GTAW process with two passes on the transition zone (ER309L filler) and three passes on the A240 TP 316L cladding layer.

Picchi et al. [63] utilised 3 different welding processes to join two butt-positioned plates of ASTM A516 GR.70 steel clad with AISI 904L austenitic steel (overall thickness 31 mm): Near the cladding layer, 1 pass of GTAW (ER70S-6 filler) as root welding was used; on the carbon steel side, 4 passes of GMAW (ER70S-6 filler) and 15 passes of SAW (F7A2-EM12K filler) were used. Finally, on the AISI 904L side, the joint was completed with electroslag strip cladding using a single pass of Inconel 625 alloy.

Recently, Ban et al. [64] carried out a comprehensive study aimed at metallurgically and mechanically characterising the welds between two bimetallic plates consisting of Q355 structural steel (10 mm thick substrate) and AISI 316 stainless steel (2 mm thick cladding). All butt-welding trials were performed by GTAW with pure Ar as the shielding gas. The use of ER316L filler only was compared with that of different fillers, such as ER50-6 for the structural steel and ER316L for the austenitic steel, which was also experimented with to fill a transition layer. The results of the mechanical tests showed that the use of different fillers allows one to meet the specification requirements and achieve excellent mechanical properties.

3.2. *Laser and Hybrid Laser/Arc Processes*

The traditional arc welding procedures require that base and clad steels are welded separately with different filler materials, to avoid undesirable dilution. Since multiple passes are required to weld clad steels, the overall efficiency of the process is limited and therefore it was deemed worth investigating the possibility of reducing the number of passes using high-penetration LBW.

In [7], the authors butt-welded two 2205/X65 clade plates, 4 mm thick, using a fibre laser apparatus in a single pass. They obtained a full-penetration weld without any defects; however, the FZ showed a non-uniform composition and a martensitic microstructure with high hardness at the level of the carbon steel. To overcome this shortcoming, in [65] two plates of ASTM A515 grade 60 structural steel clad with AISI 304 austenitic stainless steel (total thickness 9 mm) were butt-welded in an experiment with the use of both a laser beam apparatus and a hybrid laser beam–electric arc setup. Specifically, the laser beam preceded the arc; consequently, two distinct impingement points (with a distance between each other of 55 mm) and two melting pools were generated.

In the case of the laser beam alone, the filler metal was added as consumable inserts of AWS ER310 in the form of two strips (each one 0.5 mm thick), which were interposed between the butt-positioned plates (Figure 12a), whereas an ER 308 filler wire was used for the hybrid LBW-GMAW combination. In this case, on the AISI 304 side, the plates were bevelled with a V-groove, edges inclined at $\alpha = 45°$ and no gap, as shown in Figure 12b. In both cases, the plates were welded with the clad steel side exposed at the thermal source. In the LBW process, uniform composition and microhardness values were obtained along the longitudinal axis of the FZ. The hybrid process had advantageous results, achieving both deep penetration (due to the laser beam) and better tolerance for geometric defects (due to GMAW).

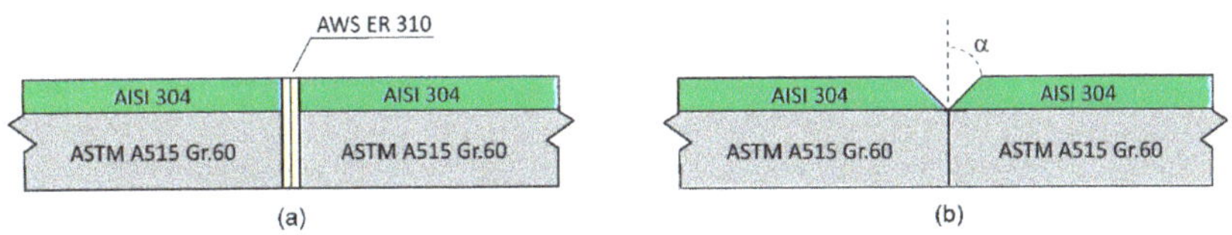

Figure 12. Preparation of the butt-positioned plates: (**a**) interposed consumable inserts for LBW; (**b**) chamfered V-groove for the hybrid LBW–GMAW combination [65].

In [66], a single-pass laser/arc hybrid welding of two AISI 304 stainless steel/Q235B carbon steel plates (total thickness 9 mm) was carried out using ER310 austenitic stainless steel as filler wire. The laser beam and torch were both placed on the clad steel side (Figure 13); in this case as well, the plates were butt-positioned; however, unlike in the previous case, they were prepared with squared edges, and the torch preceded, at a distance of 3 mm, the laser beam. In this work, the authors verified that the laser power value had greater impact on the weld corrosion resistance than that of the wire filling rate.

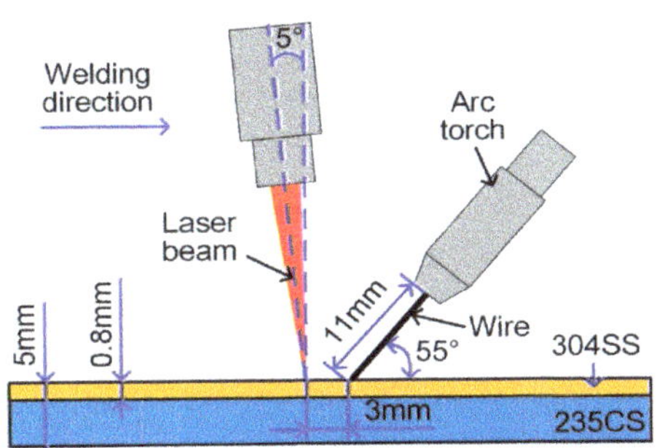

Figure 13. Setup of the hybrid LBW–GMAW process. Reproduced from [66] with permission from Springer Nature.

More recently, some authors proved that the hybrid laser–electric arc process allows welding of thick steel plates (for example, 20 [67] or 25 mm [68]). This method combines the advantages of the large penetration capacity of LBW and the strong adaptability of arc welding to any misalignments of the plates [69], with the results also being useful in enhancing the beneficial effects of laser oscillations on mechanical properties [70].

The considerations on the different welding processes carried out in this section are summarized in qualitative form in Table 1.

Table 1. Characteristics of the welding processes.

Welding Process	Welding Procedure	Welding Sequence	Filler Geometry	Weld Cross-Section Shape	Dilution Rate
Arc welding	Multi-pass	Starting from base metal	Wire	Wide	High
LBW	Single pass	-	Wire/strip	Narrow	Low
Hybrid (laser/arc)	Single pass	-	Wire/strip	Very wide at arc side	Low at laser side/high at arc side

4. Discussion

The study of clad steel welds is quite complex, since usually several passes of arc welding are carried out with different fillers for base steel and the cladding layer, as shown in the previous section. Furthermore, in fusion welding processes, such as weld overlay or welding between two clad steel plates, dilution plays a fundamental role in determining the molten pool composition, on which depend the joint properties.

The following relationships give the dilution ratios in the weld, with respect to the filler metal (d_F) and one or possibly two metals (d_{M1} and d_{M2}) that take part in the fusion process:

$$d_F = A_F/(A_F + A_{M1} + A_{M2}), \quad (1)$$

$$d_{Mi} = A_{Mi}/(A_F + A_{M1} + A_{M2}), \quad (2)$$

where A_F, A_{M1} and A_{M2} represent the volumetric contributions to the weld of the filler and the two base metals; d_{Mi} and A_{Mi} represent the dilution ratio and the volumetric contribution to the weld of one of the base metals, respectively. These quantities can be reduced by one dimension into area terms under the assumption that the cross-sectional areas do not vary along the weld bead length [71].

The dilution rate varies for each welding process. However, since for the same welding process the dilution rate can vary depending on working parameters (such as power, travel speed, preheating or inter-pass temperature), it is necessary to take each specific case into consideration. By way of example, Table 2 collects, for some processes of weld overlaying, the dilution values expressed as percentages. The wide span of the reported quantities is due to variability in the process parameters. In the same table, slightly more accurate ranges of dilution values are reported in the case of cladding with a high Ni alloy, expressed as Fe (%) diffused from the base steel, specifying the corresponding thicknesses of the cladding layers obtained with a single pass.

Table 2. Dilution, Fe (%) diffused from base steel and cladding layer thickness for some weld overlay processes.

Weld Overlay Process	Dilution (%) [40]	Fe (%)/Cladding Layer Thickness (mm) [72]
SAW	30–70	-
GMAW	13–33	13–26.9/3–3.5
SMAW	13–20	22–23.4/3–4
GTAW	13–20	12.6/-
GTAW-HW	10–40	6.6–23.7/1.6–3.5
CMT	-	1–4.1/3

As for the LBW overlay process proposed in [53], austenitic steel plates were remelted onto the carbon steel substrate, which contribute, by melting in turn, to the composition of the final layer. This process was compared to cold metal transfer (CMT), based on adding a traditional filler wire through GMAW using a pulsing action instead of continuous power. In this way, the heat input in CMT is lower than that in conventional GMAW, offering a significant benefit due to the reduced thermal effects [73].

The two processes carried out in [53] (LBW and CMT) showed the same dilution: a low value for the base metal (d_M less than 20%) and a greater one for the filler (d_F greater than 80%). When these values are known, the layer final composition can be calculated by weighted mass balance.

The resulting microstructure can be estimated through the Schaeffler diagram in Figure 14, based on equivalent compositions expressed as percentages by weight. The representative point of the layer obtained by CMT or the laser beam process indicates that the microstructure tends to be almost fully austenitic with 2–3% of residual δ-ferrite for both the processes, whereas SEM-EDS measurements showed that the two processes have distinct representative points with a ferrite content equal to about 10%, a value close to that estimated from the metallographic observations [53]. This discrepancy can be ascribed to the effect of the solidification rate, which is not considered in the Schaeffler diagram.

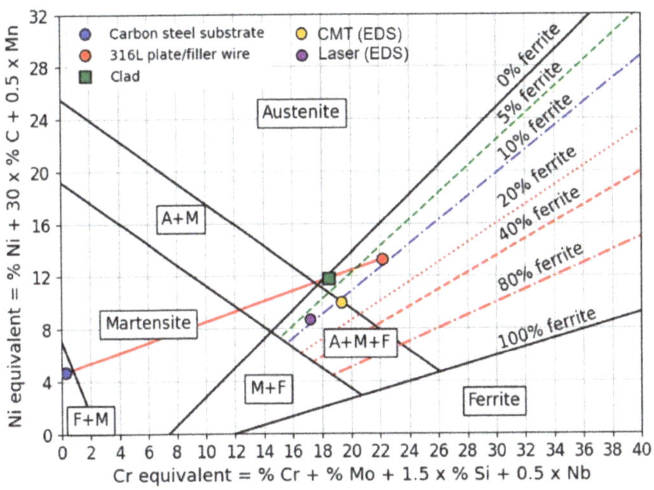

Figure 14. Schaeffler diagram with the representative points of the carbon steel substrate, AISI 316 filler (plate or wire) and clad layer (resulting from a mass balance for a carbon steel dilution d_M = 17%). The experimental compositions of the clad layer, obtained with SEM-EDS measurements, for CMT and the laser beam process are given too. Reproduced from [53].

Dilution is responsible for fusion zone composition and consequently for the weld metallurgical and mechanical features. This topic has been widely addressed in the literature, and many studies have been specifically aimed at investigating the solidification modes of the weld metal (see, for reference, [74]).

In general, the composition range of austenitic steels is broad enough that four types of solidification modes and subsequent solid-state transformations can occur, as reassumed in Table 3 in agreement with what has been reported by several authors [75,76]. To facilitate reading, the composition limits for each solidification mode have been expressed as a function of the Cr_{eq}/Ni_{eq} ratio, in which Cr_{eq} and Ni_{eq} can have the same expressions as those utilised for the Schaeffler diagram [77]. Moreover, Table 3 shows the final microstructures resulting from each solidification mode.

Table 3. Solidification modes, phase transformations, composition limits and corresponding microstructures.

Solidification Modes	Transformations	Composition Limits (Cr_{eq}/Ni_{eq})	Microstructures
Austenitic (A mode)	$L \rightarrow L + \gamma \rightarrow \gamma$	<1.25	Fully austenitic
Austenitic/ferritic (AF mode)	$L \rightarrow L + \gamma \rightarrow L + \delta + \gamma \rightarrow \gamma + \delta$	1.25–1.48	Austenitic with ferrite at cell and dendrite boundaries
Ferritic/austenitic (FA mode)	$L \rightarrow L + \delta \rightarrow L + \delta + \gamma \rightarrow \delta + \gamma$	1.48–1.95	Skeletal and/or lathy ferrite from F-A solid-state transformation
Ferritic (F mode)	$L \rightarrow L + \delta \rightarrow \delta + \gamma$ [1]	>1.95	Ferrite matrix and Widmastänstatten austenite

[1] A high Cr content leads to a full ferritic final microstructure. As temperature decreases, ferrite may transform into the brittle σ phase.

It is important to note that the A and AF solidification modes are associated with primary austenite solidification, while the FA and F types have δ-ferrite as the primary phase. The presence of this phase in the final stage of the FA solidification mode will be discussed in the following, highlighting how it provides the best resistance to solidification cracking.

For the composition range in which solidification occurs following the FA mode, the Welding Research Council (WRC) diagram (version revised in 1992) provides detailed results (according to $Cr_{eq} = Cr + Mo + 0.7{\cdot}Nb$ and $Ni_{eq} = Ni + 35{\cdot}C + 20{\cdot}N + 0.25{\cdot}Cu$) for determining the ferrite percentage of the final microstructure and is therefore currently utilised in the literature [78–80].

To highlight the different microstructures due to the physical transformations that occur during solidification and subsequent cooling, several authors [81–83] considered in their studies the Fe-Cr-Ni pseudo-binary diagram with 70% iron by weight, which can be represented according to the Cr_{eq}/Ni_{eq} ratio, as shown in Figure 15 where the limits for the FA mode are indicated by the red rectangle.

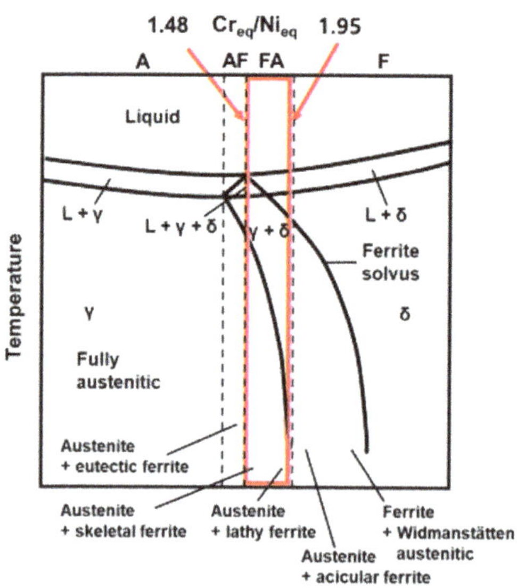

Figure 15. Fe–Cr-Ni pseudo-binary diagram (70% of iron) with indication of the FA mode composition range. Reproduced from [82].

The different expressions of Cr_{eq} and Ni_{eq}, adopted for the Schaeffler and WRC 1992 diagrams, produce small deviations in the prediction of the phases, as shown in the work by Kianersi et al. [84]. In any case, the predictions made by the aforementioned diagram are not exhaustive, since the $\delta \rightarrow \gamma$ solid-state transformation is a diffusion-controlled process depending on the cooling rate [85]. As a matter of fact, when the welding heat input increases, the cooling rate is reduced [86]; consequently, the duration of the $\delta \rightarrow \gamma$ transformation is prolonged, leading to the reduction in the final amount of residual δ-ferrite. Conversely, a lower heat input leads to faster cooling, reduced time for the $\delta \rightarrow \gamma$ transformation and therefore a greater amount of residual δ-ferrite [87]. In this regard, numerical and analytical models, as well as experimental methods, were recently developed to simulate the effects of welding parameters on thermal fields and cooling rates [88–90], as well as on residual stress [91,92], although there is a lack of articles in the literature that deal specifically with welding of clad steel plates. In any case, the cooling rate is a critical parameter as it affects the morphology of ferrite: skeletal ferrite is favoured by a low value (high heat input) while lathy columnar ferrite is favoured by a higher value (lower heat input) [93]. In particular, these morphologies can coexist in the same weld (Figure 16), with one or the other prevailing depending on the cooling rate [94].

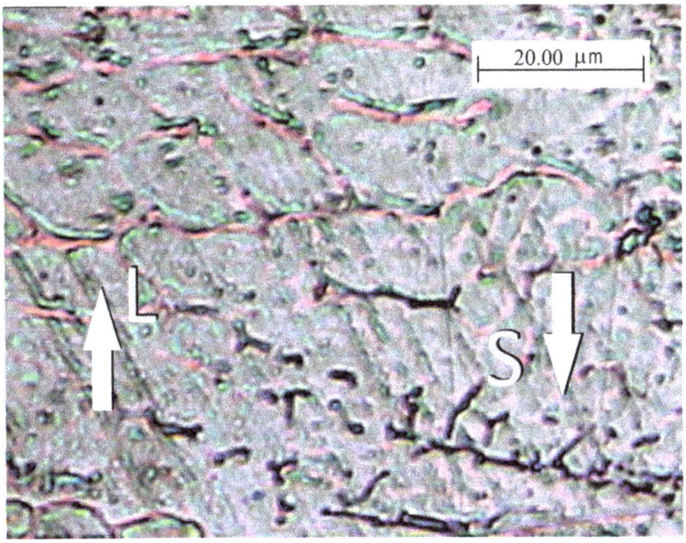

Figure 16. Detail of a fused zone (Cr_{eq}/Ni_{eq} = 1.63) with indication by the white arrows of the skeletal and lathy residual ferrite. Reproduced from [94].

The formation of weld solidification cracking depends on the solidification mode and therefore is a function of composition: it has been ascertained that the FA mode offers the greatest resistance to cracking in the finale stage of solidification, whereas the F mode is more susceptible to this phenomenon, but its behaviour is better than those of the A and AF ones [74]. In this regard, it is should be noted that, for austenitic compositions, the weld susceptibility to solidification cracking is affected by the grain boundary morphology, since primary austenite solidification (AF mode) generates straight boundaries that are easy to follow by cracks during their growth; conversely, the tortuous ferritic boundaries that characterise the FA mode produce some beneficial effects, making the growth of solidification cracks difficult [95] and also preventing hydrogen embrittlement [87].

In any case, the concentration limit of 10% for residual ferrite should not be exceeded because its excessive presence significantly reduces toughness (especially at cryogenic temperatures), deteriorates corrosion resistance and favours the formation of brittle intermetallic phases at high service temperatures [78]. Indeed, when working temperatures are in the 550–900 °C range, primary ferrite can decompose into carbides and phases detrimental for mechanical properties, such as ductility, impact toughness and creep strength as well as corrosion resistance [96].

Hybrid arc welding processes in multiple passes with the use of different filler metals could give rise to dissimilar microstructures; in general, the microstructure of the layers close to the base metal is ferritic/pearlitic, while close to the cladding steel, where fillers rich in Ni are used, it is austenitic with residual ferrite (see, for example, Figure 4 of the article by Ghorbel et al. [58]).

The presence of martensite, considered a welding defect since it tends to trigger cracks under fatigue, can be observed in the ferritic layers deposited on the austenitic ones. It forms due to dilution, which enhances the presence of alloy elements, such as Cr and Ni, and consequently causes the shift of the CCC curve and the decrease in the critical martensitic cooling rate [60]. Similar cases of martensite formation were reported in [61,62]. In [63], the formation of martensite is ascribed to values of dilution greater than 60%, which lead to the FZ in the A + M field of the Schaeffler diagram (Figure 14).

When welding using a filler metal with a composition dissimilar to that of the base metal, the diluted composition of the weld can be assumed homogeneous, due to strong hydrodynamic mixing forces in the molten pool. However, along the fusion line, the formation of very thin PDZ with an intermediate composition between the base metal and the bulk weld metal is usually observed [97].

As documented by some authors [98,99], a carbon-enriched PDZ, characterised by a hard martensitic microstructure, occurs when carbon steel base material is weld overlaid by high Ni alloy filler metal. In Figure 17, the AST A36 carbon steel base metal (BM), the AWS ER NiCrMo-3 (Inconel 625) filler metal (FM) deposited by GMAW and the interposed PDZ are outlined by the results of the Vickers microhardness test; in this case, the narrow peak around 380 HV testifies to the presence of a martensitic microstructure in the PDZ.

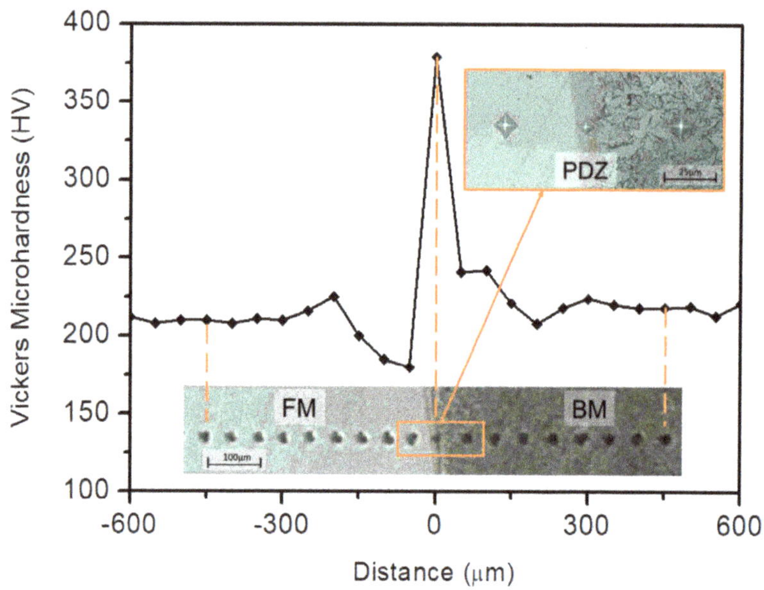

Figure 17. Vickers microhardness profile along a transversal line across the interface, from BM (on the left) to FM (on the right). Reproduced from [99].

Another issue to be considered concerns the presence of decarburization and carbon accumulation zones at the melting interface between the layers deposited with different compositions at the level of the carbon steel/austenitic steel transition zone, while martensite is formed in the substrate by diffusion of the alloying elements, more markedly as the welding heat input increases [64].

In addition to the effect on martensitic transformation, carbon accumulation generates conditions favourable to Cr-carbide precipitation. It results in a local hardness increase, which weakens the microstructure, as well as in sensitization to intergranular corrosion (see [100] for a review). In this regard, Figure 18 shows the temperature–time sensitization curves [101], and the red dashed line indicates the effect of an increase in carbon percentage; it can be deduced that the precipitation time is significantly reduced when the carbon content increases from 0.03%, which is the typical value of the AISI 304L composition, to higher values, making the precipitation of carbides possible.

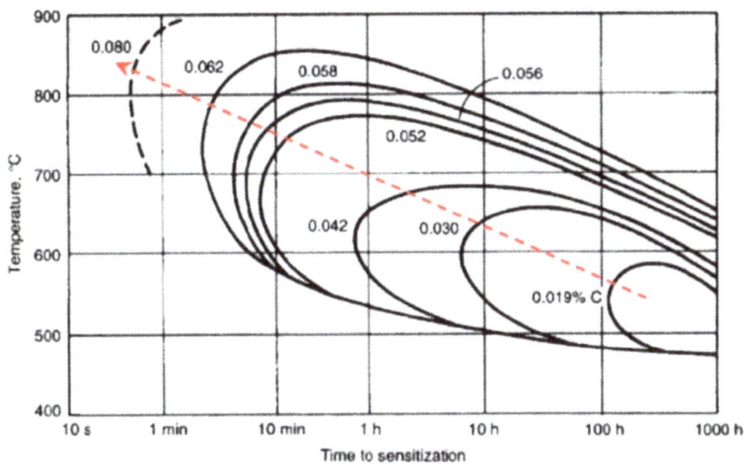

Figure 18. Precipitation curves of $Cr_{23}C_6$ carbide as a function of carbon content. The dashed red line indicates how the precipitation kinetics become faster as the carbon content increases. Reproduced from [101].

It can be deduced that the occurrence of sensitization is possible in adjacent layers where dilution may result in an increase in carbon content, Therefore, to avoid a long permanence of the weld within the critical range of temperature, a cooling time should be allowed after each pass, and the subsequent pass should initiate only when the highest temperature of the weld reaches 150 °C [59].

At the austenitic layer side, away from the cladding line, the HAZ does not undergo conditions favourable for carbide precipitation, requiring, e.g., for C = 0.03%, a permanence of approximately 8 h at 600 °C or 30 h at 500 °C (Figure 18). Such long time intervals can be excluded by means of the simulation carried out in [90] on the thermal field arising during LBW of thick AISI 304 plates.

At the base material side, carbon steel shows the characteristic transformations that occur in the HAZ. By way of example, in Figure 19 four regions are highlighted based on the temperatures reached [45], namely coarse-grained (CGHAZ), fine-grained (FGHAZ), intercritical (ICHAZ) and subcritical heat-affected zones (SCHAZ), as well as the unaffected zone. They can present martensitic and bainitic phases where sufficiently high cooling rates are reached during the welding process.

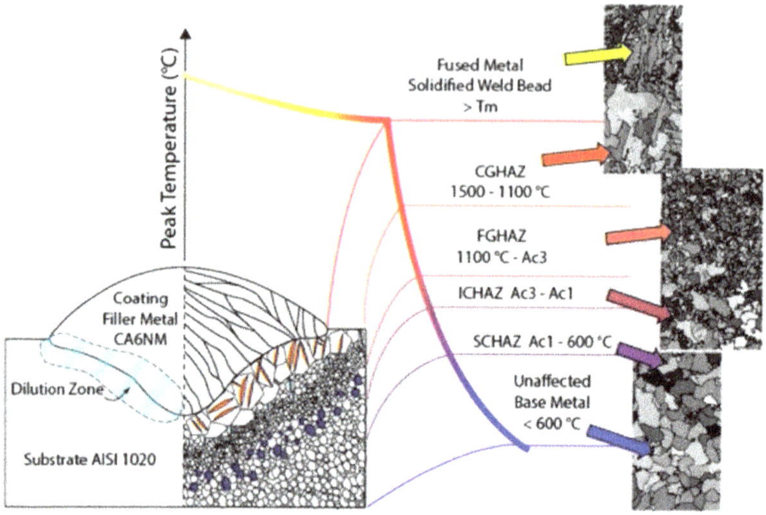

Figure 19. Schematic representation of welded microstructures and their metallurgical zones, divided according to the peak temperature. Images of the generated microstructure are displayed accordingly. Reproduced from [45].

5. Conclusions and Future Directions

This article reviewed the scientific literature on the production processes of clad steels and in particular on fusion welding of ferritic/austenitic plates. The welding processes based on multiple passes of traditional electric arc procedures are currently carried out in the petrochemical and energy industry. Alongside them, the most modern laser beam or hybrid arc–laser welding processes are increasingly used, as they allow the joining of thick plates in a single pass, which would not be possible with the usual techniques.

Over the last few decades, researchers' interest has been focused on the typical metallurgical issues that characterise the behaviour of ferritic and stainless steels. Progress in this field has developed over the last 20 years, as highlighted by the recent scientific literature. The majority of the articles examined (76/101) date back to the last five years, demonstrating the current interest in this topic.

Some issues of the production processes concern diffusion at the clad steel/base steel interface in hot rolling, formation of intermetallic compounds at the Al/steel interface in explosive cladding and dilution at the deposited cladding layer/base steel interface in weld overlay.

As for fusion welding, some issues requiring attention from researchers have emerged in this review:

- Dilution between the filler and base metal at the level of the base and clad steel, as well as dilution between the deposited layers in multi-pass welding;
- Unexpected composition of the welds;
- Solidification in non-optimal mode to ensure crack-free welds;
- Carbide enrichment in the ferritic phase, followed by martensitic transformation during cooling;
- Carbide precipitation in the austenitic phase, followed by local hardening and sensitization to intergranular corrosion as a function of the time permanence within the critical interval of temperature.

High-penetration LBW allows joining of clad steel in a single pass, even if some attention is required in the setup of the filler material to prevent dilution between the base metal and the cladding metal from causing unwanted effects.

Future work should be aimed at verifying the possibility of industrially adopting solutions capable of overcoming these shortcomings, such as the use of consumable inserts for single-pass laser welding. Moreover, further investigations are still needed on the choice of arc welding systems and filler metals suitable for clad and base steel, respectively, as well as on the edge preparation, the sequence of the multiple passes, the thermal cycle effects on the weld metallurgy and the possible occurrence of distortions or residual stresses. In this regard, it is also desirable that simulation methods are developed for the prediction of thermal fields capable of taking into account the complexity of the phenomena involved in clad steel welding.

Author Contributions: Conceptualization, F.G., S.M., C.S. and A.S.; methodology, F.G., S.M., C.S. and A.S.; investigation, F.G., S.M., C.S. and A.S.; data curation, F.G., S.M., C.S. and A.S.; writing—original draft preparation, F.G., S.M., C.S. and A.S.; writing—review and editing, F.G., S.M., C.S. and A.S.; visualization, F.G., S.M., C.S. and A.S.; supervision, F.G., S.M. and A.S. All authors have read and agreed to the published version of the manuscript.

Funding: This research received no external funding.

Data Availability Statement: No new data were created; therefore, data sharing is not applicable to this article.

Conflicts of Interest: The authors declare no conflicts of interest.

References

1. Wang, D.; Sun, X.; Jiang, Y.; Chang, X.; Yonglei, X. Review on the application of stainless-clad bimetallic steel in the marine environment. *Anti-Corros. Methods Mater.* **2024**, *71*, 132–142. [CrossRef]
2. Liu, B.X.; Wang, S.; Fang, W.; Yin, F.X.; Chen, C.X. Meso and microscale clad interface characteristics of hot-rolled stainless steel clad plate. *Mater. Charact.* **2019**, *148*, 17–25. [CrossRef]
3. Campanella, D.; Buffa, G.; Fratini, L. A two steps Lagrangian–Eulerian numerical model for the simulation of explosive welding of three dissimilar materials joints. *CIRP J. Manuf. Sci. Technol.* **2021**, *35*, 541–549. [CrossRef]
4. Dhib, Z.; Guermazi, N.; Ktari, A.; Gasperini, M.; Haddar, N. Mechanical bonding properties and interfacial morphologies of austenitic stainless steel clad plates. *Mater. Sci. Eng. A* **2017**, *696*, 374–386. [CrossRef]
5. Li, H.; Zhang, L.; Zhang, B.; Zhang, Q. Microstructure characterization and mechanical properties of stainless steel clad plate. *Materials* **2019**, *12*, 509. [CrossRef]
6. Zhu, M.; Wu, W.; Qian, W.; Xia, L.; Zhang, Y.; Wang, B. A brief review on welding of stainless steel clad plates: Issues and future perspectives. *Int. J. Adv. Manuf. Technol.* **2021**, *115*, 49–59. [CrossRef]
7. Gou, N.N.; Zhang, J.X.; Zhang, L.J.; Li, Z.G.; Bi, Z.Y. Single pass fiber laser butt welding of explosively welded 2205/X65 bimetallic sheets and study on the properties of the welded joint. *Int. J. Adv. Manuf. Technol.* **2016**, *86*, 2539–2549. [CrossRef]
8. Jiang, J.; Ding, H.; Lu, Z.; Xie, G. Interfacial microstructure and mechanical properties of stainless steel clad plate prepared by vacuum hot rolling. *J. Iron Steel Res. Int.* **2018**, *25*, 732–738. [CrossRef]
9. Jin, H.R.; Wei, R.; Wang, Y.H.; Yi, Y.L.; Jia, C.Z.; Zhao, D.X. Vacuum hot rolling preparation of stainless steel clad plate and its numerical simulation. *Strength Mater.* **2022**, *54*, 144–153. [CrossRef]
10. Dhib, Z.; Guermazi, N.; Gasperini, M.; Haddar, N. Cladding of low-carbon steel to austenitic stainless steel by hot-roll bonding: Microstructure and mechanical properties before and after welding. *Mater. Sci. Eng. A* **2016**, *656*, 130–141. [CrossRef]
11. Li, H.; Zhang, L.; Zhang, B.; Zhang, Q. Interfacial fracture behavior of a stainless/carbon steel bimetal plate in a uniaxial tension test. *Results Phys.* **2019**, *14*, 102470. [CrossRef]
12. Zhu, Z.; He, Y.; Zhang, X.; Liu, H.; Li, X. Effect of interface oxides on shear properties of hot-rolled stainless steel clad plate. *Mater. Sci. Eng. A* **2016**, *669*, 344–349. [CrossRef]
13. Jing, Y.; Qin, Y.; Zang, X.; Shang, Q.; Song, H. A novel reduction-bonding process to fabricate stainless steel clad plate. *J. Alloys Compd.* **2014**, *617*, 688–698. [CrossRef]
14. Yu, T.; Jing, Y.; Yan, X.; Li, W.; Pang, Q.; Jin, G. Microstructures and properties of roll-bonded stainless/medium carbon steel clad plates. *J. Mater. Process. Technol.* **2019**, *266*, 264–273. [CrossRef]
15. Chen, C.X.; Liu, M.Y.; Liu, B.X.; Yin, F.X.; Dong, Y.C.; Zhang, X.; Zhang, F.Y.; Zhang, Y.G. Tensile shear sample design and interfacial shear strength of stainless steel clad plate. *Fusion Eng. Des.* **2017**, *125*, 431–441. [CrossRef]
16. Yang, Y.; Jiang, Z.; Liu, X.; Sun, J.; Wang, W. Enhanced interfacial strength and ductility of stainless steel/carbon steel laminated composite by heterogenous lamella structure. *J. Mater. Res. Technol.* **2022**, *18*, 4846–4858. [CrossRef]
17. Dai, P.; Li, S.; Wu, L.; Wang, Y.; Feng, G.; Deng, D. A new numerical model to predict welding-induced sensitization in SUS304 austenitic stainless steel joint. *J. Mater. Res. Technol.* **2022**, *17*, 234–243. [CrossRef]
18. Giudice, F.; Missori, S.; Murdolo, F.; Sili, A. Metallurgical characterization of the interfaces in steel plates clad with austenitic steel or high Ni alloys by hot rolling. *Metals* **2020**, *10*, 286. [CrossRef]

19. *ASTM A262-15*; Standard Practices for Detecting Susceptibility to Intergranular Attack in Austenitic Stainless Steels. ASTM International: West Conshohocken, PA, USA, 2021.
20. Liu, B.X.; Yin, F.X.; Dai, X.L.; He, J.N.; Fang, W.; Chen, C.X.; Dong, Y.C. The tensile behaviors and fracture characteristics of stainless steel clad plates with different interfacial status. *Mater. Sci. Eng. A* **2017**, *679*, 172–182. [CrossRef]
21. Li, Q.; Chen, W.; Du, J.; Lu, M.; Wang, Z.; Huang, Y. Microstructure and coordination mechanism of interface of stainless steel/carbon steel cladding plate prepared by vacuum diffusion bonding. *Mater. Sci. Eng. A* **2022**, *829*, 142178. [CrossRef]
22. Song, H.; Shin, H.; Shin, Y. Heat-treatment of clad steel plate for application of hull structure. *Ocean Eng.* **2016**, *122*, 278–287. [CrossRef]
23. Wang, H.; Wang, Y. High-velocity impact welding process: A review. *Metals* **2019**, *9*, 144. [CrossRef]
24. Sherpa, B.B.; Rani, R. Advancements in explosive welding process for bimetallic material joining: A review. *J. Alloys Met. Syst.* **2024**, *6*, 100078. [CrossRef]
25. Kaur, J.; Mangla, V.; Singh, J.; Kumar, S.; Srivastava, N. Cladding of stainless steel (SS304) on aluminium alloy (AA1100) by explosive welding. *Mater. Today Proc.* **2018**, *5*, 19136–19139. [CrossRef]
26. Meco, S.; Ganguly, S.; Williams, S.; McPherson, N. Design of laser welding applied to T joints between steel and aluminium. *J. Mater. Process. Technol.* **2019**, *268*, 132–139. [CrossRef]
27. Carvalho, G.H.S.F.L.; Galvão, I.; Mendes, R.; Leal, R.M.; Loureiro, A. Aluminum-to-steel cladding by explosive welding. *Metals* **2020**, *10*, 1062. [CrossRef]
28. Carvalho, G.H.S.F.L.; Galvão, I.; Mendes, R.; Leal, R.M.; Loureiro, A. Formation of intermetallic structures at the interface of steel-to-aluminium explosive welds. *Mater. Charact.* **2018**, *142*, 432–442. [CrossRef]
29. Yang, M.; Ma, H.; Shen, Z.; Sun, Y. Study on explosive welding for manufacturing meshing bonding interface of CuCrZr to 316L stainless steel. *Fusion Eng. Des.* **2019**, *143*, 106–114. [CrossRef]
30. Shi, Y.; Wang, J.; Zhou, X.; Xue, X.; Li, H. Post-fire constitutive model on explosively welded stainless-clad bimetallic steel after cold-forming process. *J. Constr. Steel Res.* **2023**, *209*, 108038. [CrossRef]
31. Carvalho, G.H.S.F.L.; Galvão, I.; Mendes, R.; Leal, R.M.; Loureiro, A. The role of physical properties in explosive welding of copper to stainless steel. *Def. Technol.* **2023**, *22*, 88–98. [CrossRef]
32. Ayele, B.S.; Mebratie, B.A.; Meku, A.A. Investigation of the effect of explosive welding parameters on aluminum–steel bimetals: A numerical approach. *J. Mater. Civil Eng.* **2023**, *35*, 11. [CrossRef]
33. Carvalho, G.H.S.F.L.; Galvão, I.; Mendes, R.; Leal, R.M.; Loureiro, A. Microstructure and mechanical behaviour of aluminium-carbon steel and aluminium-stainless steel clads produced with an aluminium interlayer. *Mater. Charact.* **2019**, *155*, 109819. [CrossRef]
34. Palaci, Y.; Olgun, M. Influences of heat treatment on mechanical behavior and microstructure of the explosively welded D grade steel/En Aw 5083 aluminium joint. *Arch. Metall. Mater.* **2021**, *66*, 373–380. [CrossRef]
35. Liu, W.; Ma, J.; Atabaki, M.M.M.; Kovacevic, R. Joining of advanced high-strength steel to AA 6061 alloy by using Fe/Al structural transition joint. *Mater. Des.* **2015**, *68*, 146–157. [CrossRef]
36. Costanza, G.; Crupi, V.; Guglielmino, E.; Sili, A.; Tata, M.E. Metallurgical characterization of an explosion welded aluminum/steel joint. *Metall. Ital.* **2016**, *11*, 17–22.
37. Lee, T.; Zhang, S.; Vivek, A.; Daehn, G.; Kinsey, B. Wave formation in impact welding: Study of the Cu–Ti system. *CIRP Ann.–Manuf. Techn.* **2019**, *68*, 261–264. [CrossRef]
38. Gonçalves e Silva, R.H.; Riffel, K.C.; da Paixão Carvalho, L.; Kejelin, N.Z. Double-sided welding as an alternative for joining internally clad pipes. *J. Pipeline Syst. Eng. Pract.* **2020**, *11*, 04020012. [CrossRef]
39. *Norsok Standard M-001*; Materials Selection, Edition 5. Standards Norway: Lysaker, Norway, 2014.
40. Volpi, A.; Serra, G. Weld overlay of highly corrosion resistant nickel chromium molybdenum alloys, UNS N06059, on low alloy equipment operating at high temperature. In Proceedings of the ASME 2018 Symposium on Elevated Temperature Application of Materials for Fossil, Nuclear, and Petrochemical Industries ETAM 2018, Seattle, WA, USA, 3–5 April 2018. [CrossRef]
41. Saha, M.K.; Hazra, R.; Mondal, A.; Das, S. Effect of heat input on geometry of austenitic stainless steel weld bead on low carbon steel. *J. Inst. Eng. India Ser. C* **2019**, *100*, 607–615. [CrossRef]
42. Aslam, M.; Sahoo, C.K. Numerical and experimental investigation for the cladding of AISI 304 stainless steel on mild steel substrate using Gas Metal Arc Welding. *CIRP J. Manuf. Sci. Technol.* **2022**, *37*, 378–387. [CrossRef]
43. Singhal, T.S.; Jain, J.K. GMAW cladding on metals to impart anti-corrosiveness: Machine, processes and materials. *Mater. Today Proc.* **2020**, *26*, 2432–2441. [CrossRef]
44. Sowrirajan, M.; Koshy Mathews, P.; Vijayan, S.; Amaladasan, Y. Effect of weld dilution on post-weld thermal conductivity of austenitic stainless steel clad layers. *Mater. Res. Express* **2018**, *5*, 096512. [CrossRef]
45. Sartori Moreno, J.; Faria Conde, F.; Alves Correa, C.; Barbosa, L.H.; Pereira da Silva, E.; Avila, J.; Buzolin, R.H.; Cavalcanti Pinto, H. Pulsed FCAW of martensitic stainless clads onto mild steel: Microstructure, hardness, and residual stresses. *Materials* **2022**, *15*, 2715. [CrossRef] [PubMed]
46. Cattivelli, A.; Roy, M.J.; Burke, M.G.; Dhers, J.; Lee, T.L.; Francis, J.A. Internal stresses in a clad pressure vessel steel during post weld heat treatment and their relevance to underclad cracking. *Int. J. Pres. Ves. Pip.* **2021**, *193*, 104448. [CrossRef]

47. Matias, J.V.S.; Lourenço, M.J.C.; Jorge, J.C.F.; de Souza, L.F.G.; Farneze, H.N.; Mendes, M.C.; da Silva, C.L.A.; Araújo, L.S. Behavior of a superaustenitic stainless steel weld cladding deposited by the gas metal arc welding process. *Mater. Today Commun.* **2023**, *34*, 104978. [CrossRef]
48. Hao, Y.; Li, J.; Li, X.; Liu, W.; Cao, G.; Li, C.; Liu, Z. Influences of cooling rates on solidification and segregation characteristics of Fe-Cr-Ni-Mo-N super austenitic stainless steel. *J. Mater. Process. Technol.* **2020**, *275*, 116326. [CrossRef]
49. Matias, J.V.S.; Lourenco, M.J.C.; Jorge, J.C.F.; de Souza, L.F.G.; Farneze, H.N.; Mendes, M.C.; Araujo, L.S. Microstructure and corrosion properties of the AISI 904L weld cladding obtained by the electro slag process. *J. Mater. Res. Technol.* **2021**, *15*, 5151–5164. [CrossRef]
50. Costa, J.F.M.; Lacerda, P.L.; Magalhães, H.A.Y.; Jorge, J.C.F.; de Souza, L.F.G.; Mendes, M.C.; Araújo, L.S.; Farneze, H.N. Inconel 625 weld claddings obtained by the GMAW-RE with rotating electrode. *Int. J. Adv. Manuf. Technol.* **2024**, *132*, 5647–5661. [CrossRef]
51. Ungethüm, T.; Spaniol, E.; Hertel, M.; Füssel, U. Analysis of metal transfer and weld geometry in hot-wire GTAW with indirect resistive heating. *Weld. World* **2020**, *64*, 2109–2117. [CrossRef]
52. Rodrigo Stohler Gonzaga, R.S.; Farias, F.W.C.; da Cruz Payão Filho, J. Microstructural characterization of the transition zone between a C–Mn steel pipe and a 70%Ni30%Cu alloy cladding welded by HW-GTAW. *Int. J. Pres. Ves. Pip.* **2021**, *192*, 104433. [CrossRef]
53. Bunaziv, I.; Ren, X.; Hagen, A.B.; Hovig, E.W.; Jevremovic, I.; Dahl, S.G. Laser beam remelting of stainless steel plate for cladding and comparison with conventional CMT process. *Int. J. Adv. Manuf. Technol.* **2023**, *127*, 911–934. [CrossRef]
54. Li, C.; Zhu, J.; Cai, Z.; Mei, L.; Jiao, X.; Du, X.; Wang, K. Microstructure and corrosion resistance of under water laser cladded duplex stainless steel coating after underwater laser remelting processing. *Materials* **2021**, *14*, 4965. [CrossRef]
55. Wang, Z.D.; Sun, G.F.; Chen, M.Z. Investigation of the underwater laser directed energy deposition technique for the on-site repair of HSLA-100 steel with excellent performance. *Addit. Manuf.* **2021**, *39*, 101884. [CrossRef]
56. Sun, G.; Wang, Z.; Lu, Y.; Chen, M.; Yang, K.; Ni, Z. Underwater laser welding/cladding for high-performance repair of marine metal materials: A review. *Chin. J. Mech. Eng.* **2022**, *35*, 5. [CrossRef]
57. Ghorbel, R.; Ktari, A.; Haddar, N. Microstructure and mechanical property assessment of stainless steel–clad plate joint made by hybrid SMAW-GTAW multi-pass welding process. *Weld. World* **2022**, *66*, 1593–1608. [CrossRef]
58. Ghorbel, R.; Ktari, A.; Haddar, N. On the origin of the local hardening zone on welded stainless clad steel plates. *Fusion Sci. Technol.* **2022**, *78*, 503–511. [CrossRef]
59. Ghorbel, R.; Ktari, A.; Haddar, N. Experimental analysis of temperature field and distortions in multi-pass welding of stainless cladded steel. *Int. J. Adv. Manuf. Technol.* **2021**, *113*, 3525–3542. [CrossRef]
60. An, Q.; Fan, K.Y.; Ge, Y.F.; Liu, B.X.; Liu, Y.C.; Wang, S.; Chen, C.X.; Ji, P.G.; Yin, F.X. Microstructure and mechanical properties of stainless steel clad plate joints produced by TIG and MAG hybrid welding. *J. Adhes. Sci. Technol.* **2020**, *34*, 670–685. [CrossRef]
61. Yu, W.X.; Liu, B.X.; Chen, C.X.; Liu, M.Y.; Zhang, X.; Fang, W.; Ji, P.G.; He, J.N.; Yin, F.X. Microstructure and mechanical properties of stainless steel clad plate welding joints by different welding processes. *Sci. Technol. Weld. Join.* **2020**, *25*, 571–580. [CrossRef]
62. Li, C.; Qin, G.; Tang, Y.; Zhang, B.; Lin, S.; Geng, P. Microstructures and mechanical properties of stainless steel clad plate joint with diverse filler metals. *J. Mater. Res. Technol.* **2020**, *9*, 2522–2534. [CrossRef]
63. Bezerra de Mello Picchi, I.; Lima dos Santos, M.; Bezerra de Santana, T.H.; Costa de Santana, R.A.; Torres López, E.A.; de Albuquerque Vicente, A.; de Abreu Santos, T.F. Microstructure characterization of the weld cladding of clad steel plate A516 GR.70—AISI 904L by electroslag strip cladding using Inconel 625 strip. *Weld. World* **2023**, *67*, 2571–2587. [CrossRef]
64. Ban, H.; Yang, X.; Shi, Y.; Chung, K.-F.; Hu, Y.-F. Micro-macro properties of stainless-clad bimetallic steel welded connections with different configurations. *J. Constr. Steel Res.* **2024**, *217*, 108637. [CrossRef]
65. Missori, S.; Sili, A. Prediction of weld metal microstructure in laser beam welded clad steel. *Metallurgist* **2018**, *62*, 84–92. [CrossRef]
66. Meng, Y.; Kang, K.; Gao, M.; Zeng, X. Relationship between corrosion resistance and microstructure characteristic of single-pass laser-arc hybrid welded stainless clad steel plate. *Metall. Mater. Trans. A* **2019**, *50A*, 2817–2825. [CrossRef]
67. Brunner-Schwer, C.; Üstündag, Ö.; Bakir, N.; Gumenyuk, A.; Rethmeier, M. Process advantages of laser hybrid welding compared to conventional arc-based welding processes for joining thick steel structures of wind tower. *IOP Conf. Ser. Mater. Sci. Eng.* **2023**, *1296*, 012028. [CrossRef]
68. Zhang, L.; Peng, G.; Chi, J.; Bi, J.; Yuan, X.; Li, W.; Zhang, L. Effect of process parameters on the formability, microstructure, and mechanical properties of laser-arc hybrid welding of Q355B steel. *Materials* **2023**, *16*, 4253. [CrossRef] [PubMed]
69. Yang, S.; Yang, L.; Wang, D.; Zhang, F.; Liu, C.; Huang, G. Effect of welding stability on process porosity in laser arc hybrid welding of dissimilar steel. *Optik* **2022**, *271*, 170165. [CrossRef]
70. Hao, K.; Gao, Z.; Huang, J.; Xu, L.; Liu, Y.; Han, Y.; Zhao, L.; Ren, W. Comparisons of laser and laser-arc hybrid welded carbon steel with beam oscillation. *Opt. Laser Technol.* **2023**, *157*, 108787. [CrossRef]
71. DuPont, J.N. Dilution in fusion welding. In *Metals Handbook*, 1st ed.; Lienert, T.J., Babu, S.S., Siewert, T.A., Acoff, V.L., Eds.; ASM: Materials Park, OH, USA, 2011; Volume 06A, pp. 115–121.
72. Costa, J.F.M.; Jorge, J.C.F.; de Souza, L.F.G.; Mendes, M.C.; Farneze, H.N.; Magalhães, H.A.Y. Microstructural evaluation of Inconel 625 weld cladding deposited by the GMAW and GMAW with rotating electrode processes. *Tecnol. Metal. Mater. Min.* **2024**, *21*, e2848. [CrossRef]

73. Bellamkonda, P.N.; Dwivedy, M.; Addanki, R. Cold metal transfer technology—A review of recent research developments. *Results Eng.* **2024**, *23*, 102423. [CrossRef]
74. Lippold, J.C.; Kotecki, D.J. *Welding Metallurgy and Weldability of Stainless Steels*; Wiley & Sons, Inc.: Hoboken, NJ, USA, 2005; pp. 141–229.
75. Großwendt, F.; Becker, L.; Röttger, A.; Chehreh, A.B.; Strauch, A.L.; Uhlenwinkel, V.; Lentz, J.; Walther, F.; Fechte-Heinen, R.; Weber, S.; et al. Impact of the allowed compositional range of additively manufactured 316L stainless steel on processability and material properties. *Materials* **2021**, *14*, 4074. [CrossRef]
76. Nedjad, S.H.; Yildizb, M.; Saboorie, A. Solidification behaviour of austenitic stainless steels during welding and directed energy deposition. *Sci. Technol. Weld. Join.* **2023**, *28*, 1–17. [CrossRef]
77. Fu, J.W.; Yang, Y.S.; Guo, J.J.; Ma, J.C.; Tong, W.H. Microstructure evolution in AISI 304 stainless steel during near rapid directional solidification. *Mater. Sci. Technol.* **2009**, *25*, 1013–1016. [CrossRef]
78. Vashishtha, H.; Taiwade, R.V.; Sharma, S.; Patil, A.P. Effect of welding processes on microstructural and mechanical properties of dissimilar weldments between conventional austenitic and high nitrogen austenitic stainless steels. *J. Manuf. Process.* **2017**, *25*, 49–59. [CrossRef]
79. Alali, M.; Abass, M.H.; Abbas, W.S.; Shehabd, A.A. Effect of nickel powder buffering layer on microstructure and hardness properties of high carbon steel/stainless steel arc stud welding. *Mater. Res.* **2020**, *23*, e20190567. [CrossRef]
80. Tandon, V.; Patil, A.P.; Kowshik, S. Impact of filler electrodes on welding properties of dissimilar welded 316L/201 austenitic stainless steels. *Eng. Proc.* **2023**, *59*, 90. [CrossRef]
81. Khuenkaew, T.; Kanlayasiri, K. Resistance spot welding of SUS316L austenitic/SUS425 ferritic stainless steels: Weldment characteristics, mechanical properties, phase transformation and solidification. *Metals* **2019**, *9*, 710. [CrossRef]
82. Lee, S.H. CMT-based wire arc additive manufacturing using 316L stainless steel: Effect of heat accumulation on the multi-layer deposits. *Metals* **2020**, *10*, 278. [CrossRef]
83. Giudice, F.; Sili, A. Weld metal microstructure prediction in laser beam welding of austenitic stainless steel. *Appl. Sci.* **2021**, *11*, 1463. [CrossRef]
84. Kianersi, D.; Mostafaei, A.; Amadeh, A.A. Resistance spot welding joints of AISI 316L austenitic stainless steel sheets: Phase transformations, mechanical properties and microstructure characterizations. *Mater. Des.* **2014**, *61*, 251–263. [CrossRef]
85. Tandon, V.; Thombre, M.A.; Patil, A.P.; Taiwade, R.V.; Vashishtha, H. Effect of heat input on the microstructural, mechanical, and corrosion properties of dissimilar weldment of conventional austenitic stainless steel and low-nickel stainless steel. *Metallogr. Microstruct. Anal.* **2020**, *9*, 668–677. [CrossRef]
86. Mohammed, G.R.; Ishak, M.; Aqida, S.N.; Abdulhadi, H.A. Effects of heat input on microstructure corrosion and mechanical characteristics of welded austenitic and duplex stainless steels: A review. *Metals* **2017**, *7*, 39. [CrossRef]
87. Zhou, C.; Dia, P.; Wu, H.; He, M.; Liu, X.; Chu, P.K. Effect of the ferrite morphology on hydrogen embrittlement of MAG welded 304 austenitic stainless steel. *Appl. Surf. Sci.* **2022**, *606*, 154866. [CrossRef]
88. Sepe, R.; De Luca, A.; Greco, A.; Armentani, E. Numerical evaluation of temperature fields and residual stresses in butt weld joints and comparison with experimental measurements. *Fatigue Fract. Eng. Mater. Struct.* **2021**, *44*, 182–198. [CrossRef]
89. Kik, T. Heat source models in numerical simulations of laser welding. *Materials* **2020**, *13*, 2653. [CrossRef]
90. Giudice, F.; Missori, S.; Sili, A. Parameterized multipoint-line analytical modeling of a mobile heat source for thermal field prediction in laser beam welding. *Int. J. Adv. Manuf. Technol.* **2021**, *112*, 1339–1358. [CrossRef]
91. Nezamdost, M.R.; Nekouie Esfahani, M.R.; Hashemi, S.H.; Mirbozorgi, S.A. Investigation of temperature and residual stresses field of submerged arc welding by finite element method and experiments. *Int. J. Adv. Manuf. Technol.* **2016**, *87*, 615–624. [CrossRef]
92. Giudice, F.; Sili, A. A theoretical approach to the residual stress assessment based on thermal field evaluation in laser beam welding. *Int. J. Adv. Manuf. Technol.* **2022**, *123*, 2793–2808. [CrossRef]
93. Li, Y.; Luo, Y.; Li, J.; Song, D.; Xu, B.; Chen, X. Ferrite formation and its effect on deformation mechanism of wire arc additive manufactured 308 L stainless steel. *J. Nucl. Mater.* **2021**, *550*, 152933. [CrossRef]
94. Giudice, F.; Missori, S.; Sili, A. Dissimilar welding of thick ferritic/austenitic steels plates using two laser beams in a single pass. *J. Manuf. Mater. Process.* **2024**, *8*, 134. [CrossRef]
95. Coniglio, N.; Cross, C.E. Effect of weld travel speed on solidification cracking behavior. Part 3: Modeling. *Int. J. Adv. Manuf. Technol.* **2020**, *107*, 5039–5051. [CrossRef]
96. Wang, Q.; Chen, S.; Rong, L. Properties of heavy-section AISI 316 stainless steel casting. *Metal. Mater. Trans. A* **2020**, *51*, 2998–3008. [CrossRef]
97. Kejelin, N.Z.; de Almeida Buschinelli, A.J.; Pepe, A.M. Effect of welding parameters on the partially diluted zones formation at dissimilar metal welds. In Proceedings of the 18th International Congress of Mechanical Engineering, Ouro Preto, Brazil, 6–11 November 2005.
98. Alvarães, C.P.; Madalena, F.C.A.; Souza, L.F.G.; Jorge, J.F.C.; Araujo, L.S.; Mendes, M.C. Performance of the INCONEL 625 alloy weld overlay obtained by FCAW process. *Matéria* **2019**, *24*, e-12290. [CrossRef]
99. Silva, W.C.; Nascimento, R.M.; Nascimento, A.B.G.; Mendes, A.M.; Castro, N.A.; Silva, C.L.M. S Partially diluted zones in dissimilar cladding with AWS ER NiCrMo-3 alloy deposited by the MIG process on ASTM A36 steel: Analysis and characterization by EBSD. *Mater. Res.* **2024**, *27*, e20230440. [CrossRef]

100. Gajjar, P.K.; Khatri, B.C.; Siddhpura, A.M.; Siddhpura, M.A. Sensitization and desensitization (healing) in austenitic stainless steel: A critical review. *Trans. Indian Inst. Met.* **2022**, *75*, 1411–1427. [CrossRef]
101. Bunaziv, I.; Olden, V.; Akselsen, O.M. Metallurgical aspects in the welding of clad pipelines—A global outlook. *Appl. Sci.* **2019**, *9*, 3118. [CrossRef]

materials

MDPI

Article

The Influence of Coaxial Ultrasound on the Droplet Transfer of High Nitrogen Steel GMAW Process

Jiawen Luo, Zhizheng He, Zeng Liu, Zihuan Hua, Bin Teng and Chenglei Fan *

State Key Laboratory of Advanced Welding and Joining, Harbin Institute of Technology, Harbin 150001, China
* Correspondence: fclwh@hit.edu.cn

Abstract: The nitrogen bubble bursting phenomenon during the welding process of high nitrogen steel (HNS) can lead to unstable droplet transfer and welding process, reducing the quality of weld formation. In this study, a novel approach, ultrasonic-assisted gas metal arc welding (U-GMAW), is proposed to suppress the escape of nitrogen gas during droplet transfer. This study investigates the influence of ultrasound on the metal transfer process during two distinct metal transfer modes: short-circuiting and droplet transfer. Ultrasound has a significant effect on the welding process; as ultrasonic power increases, both the arc length and droplet size decrease, while the droplet transfer frequency increases and the electrical signal stabilizes. Under the experimental conditions of this study, ultrasound has the most effective improvement on the metal transfer behavior when the ultrasonic power reaches 2 kW. Ultrasound enhances the stability of the droplet transfer process, making U-GMAW an effective and novel approach for controlling the droplet transfer behavior of high nitrogen steel.

Keywords: high nitrogen steel; ultrasonic-assisted GMAW; droplet transfer behavior; coaxial ultrasound

1. Introduction

High nitrogen stainless steel (HNS) is a rapidly developing type of stainless steel in recent years. It utilizes nitrogen as the main austenite-forming element instead of nickel [1]. Nitrogen is a strong austenite-forming element. Dissolved nitrogen improves the strength and corrosion resistance of stainless steel [2]. HNS has promising applications in various fields, such as shipbuilding, electric power, and weapon manufacturing [3]. It is also used in the medical [4] and biological [5,6] fields due to its good compatibility with the human body. Recently, HNS welding wires have been utilized in the welding of HNS. However, the elevated nitrogen content presents challenges to fusion welding. During the Gas Metal Arc Welding (GMAW) process, the molten wire forms a droplet at the tip. Since the solubility of nitrogen in solids is much greater than that in liquids [7,8], nitrogen tends to accumulate in the molten droplet. When supersaturated nitrogen atoms aggregate to form nitrogen bubbles of a certain size, the droplet transfer becomes unstable, and the droplet may expand and splash, resulting in an unstable welding process and poor weld formation [9,10]. If the nitrogen bubbles trapped in the molten pool cannot escape before solidification, they will remain in the weld, forming gas pores that degrade joint quality. Furthermore, the loss of nitrogen content can alter the microstructure and negatively affect the mechanical properties of the joint [11,12]. Several studies have highlighted the instability of droplet transfer in the GMAW of HNS. Yang et al. [13] found the expansion, explosion, and splashing of large-sized HNS droplets captured by high-speed cameras. Liu et al. [14] found that with the increase in nitrogen content in the welding wire, the intensity of nitrogen gas escaping from the droplet gradually increases, leading to unstable arc and exploding droplets, which in turn cause spatter and smoke. Yang et al. [15] also reported that in the Wire Arc Additive Manufacturing (WAAM) process of high nitrogen austenitic stainless steel (HNASS), a sharp decrease in nitrogen solubility in the molten droplet caused violent explosions and splashing, leading to defects such as incomplete fusion, cracks, and inclusions.

A decrease in the dissolved nitrogen content in the weld joint adversely affects the mechanical properties, necessitating measures to prevent nitrogen loss. Many studies have addressed nitrogen loss by optimizing shielding gas composition and content or modifying the composition of welding wires. Cui et al. [16] found that the droplet transfer mode shifts to spray transfer when O_2 is added to the shielding gas, improving weld formation and stability. The highest nitrogen content in the weld seam is achieved when using 89%Ar + 10%N_2 + 1%O_2 as the shielding gas. Liu et al. [14] designed three high nitrogen steel welding wires for GMAW. As the nitrogen content of the welding wire increased, nitrogen loss in the weld became more serious, and observation of the droplet transfer showed obvious growth and rupture upon droplet detachment. Ultrasonic vibration has also been shown to reduce nitrogen gas pores. Cui et al. [17] studied ultrasonic-assisted laser-arc hybrid welding of HNS. The porosity defects are suppressed under an ultrasonic power of 180 W, resulting in optimal mechanical properties. However, the nitrogen content decreases when the ultrasonic power exceeds a certain threshold, leading to a decrease in the hardness of the weld seam.

Research on the control of nitrogen content loss mainly focuses on the design of the components of shielding gas and welding wire to add nitrogen to the molten pool, while few studies on external auxiliary methods have been studied. Droplet transfer is a critical stage of GMAW, during which nitrogen bubbles are generated and ruptured. Droplet transfer behavior directly impacts welding quality, process stability, and efficiency of GMAW. Therefore, it is necessary to take measures to control the droplet transfer behavior of GMAW for high nitrogen steel, inhibit the growth of nitrogen bubbles, and thus reduce nitrogen loss. According to our previous research [18], ultrasound significantly compresses the arc, making it brighter compared to conventional GMAW. Additionally, the droplet transfer frequency and stability of the molten droplet increase under the influence of the acoustic radiation force.

This study aims to explore a novel approach to coaxial ultrasonic assistance for controlling the droplet transfer behavior in the GMAW process of HNS welding wire, with the goal of mitigating nitrogen loss. The ultrasonic probe vibrates coaxially with the welding wire, and the direction of ultrasonic vibration aligns with the transfer direction of the molten droplet. The influence of ultrasonic power on two typical droplet transfer modes was investigated by analyzing the droplet transfer behavior captured by a high-speed camera and the corresponding electrical signals. Subsequently, the mechanisms by which ultrasound influences these two droplet transfer modes were analyzed, providing a new method for controlling the droplet transfer behavior in HNS GMAW. By introducing ultrasonic assistance to the GMAW process, this study presents a unique perspective and practical solution for enhancing the stability of HNS droplet transfer behavior. It provides a strong guarantee for the safe and reliable operation of engineering welding structures and promotes the broader application of HNS in various industries.

2. Materials and Methods

The base material is 304 stainless steel, with dimensions of 130 mm × 50 mm × 7 mm, and the welding wire has a diameter of 1.2 mm, with nitrogen content of 0.35%, as shown in Table 1. The shielding gas composition is 98%Ar + 2%O_2, and the gas flow rate is 20 L/min. The substrate surface was polished and cleaned to degrease with ethanol before deposition.

Table 1. Composition of High Nitrogen Steel Welding Wire (wt.%).

Element	C	Si	Mn	Cr	Ni	Mo	N
Content	0.071	0.832	8.84	22.26	6.54	0.27	0.35

The ultrasonic-assisted GMAW (U-GMAW) platform consists of a welding system and a high-speed camera acquisition system, as illustrated in Figure 1. The ultrasonic transducer is coaxially integrated with the welding wire, and both the *z*-axis of the ultrasonic equipment and the GMAW welding gun can be independently adjusted. The maximum power output of the ultrasonic transducer is 2 kW. The experiments were conducted with the CLOOS QinTron welding machine (CLOOS, Haiger, Germany) operating on direct current. A high-speed camera captures images of the whole process of droplet formation and its transfer into the molten pool, with a photo resolution of 512×512 dpi, a capture frequency of 2000 fps, and exposure settings optimized for the specific experimental conditions. A filter with a wavelength of 808 nm and bandwidth of ± 10 nm is placed in front of the camera to improve image quality. The high-speed video is subsequently played back in slow motion to analyze the details of the droplet transfer and analyze the droplet and arc behavior to determine the droplet transfer mode. The high-speed imaging system used is a PHANTOM v311 (Vision Research, Wayne, NJ, USA). Prior to welding, a copper plate is placed on the workbench to facilitate heat dissipation. The high nitrogen steel substrate is fixed on the copper plate and workbench using a fixture, allowing it to move along the slider at a controlled speed.

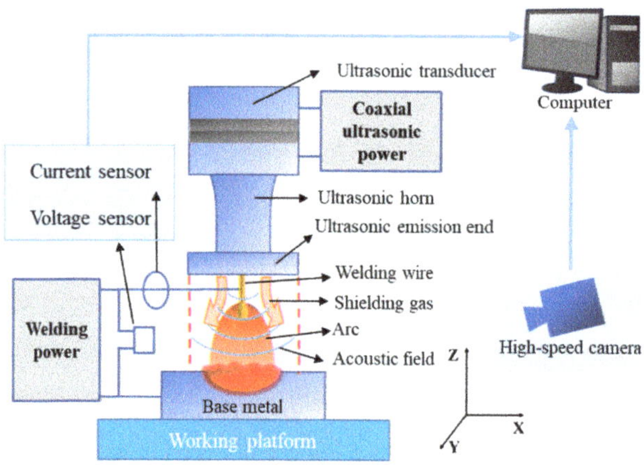

Figure 1. U-GMAW platform.

The distribution of droplet transfers of the HNS welding wire was analyzed. The welding parameters are presented in Table 2. Welding Voltages (U) of 22, 24.5, 26, 28.5 V, and 30 V, and Wire Feed Speeds (WFS) of 1.5, 3.5, 6, 8, and 10.5 m/min were selected. The high-speed camera captured the droplet transfer behavior of each experimental group and recorded their transfer modes. The main ultrasonic process parameters include the ultrasonic emission height (H, mm) and the ultrasonic power (P, kW). During the short circuit transfer, the ultrasonic emission height is set at the First Resonant Height (H_1), where the ultrasonic effect is strongest. According to the preliminary experiment, H_1 was measured as 14 mm. In the case of globular transfer, where the arc voltage is higher and the arc is longer, the ultrasonic emitter operates at the Second Resonant Height (H_2) of 20 mm to avoid damage to the emitter and to maximize the ultrasonic radiation force.

Table 2. Welding parameters.

Welding Current (I)/A	Welding Voltage (U)/V	Wire Feeding Speed (WFS)/($\mathrm{m \cdot min^{-1}}$)
120~350	20~32	1.5~12

3. Results

3.1. Distribution of Droplet Transfer Modes

The distribution of droplet transfer modes under different welding parameters is presented in Figure 2. Four droplet transfer modes are observed: short circuit transfer (red area), globular transfer (green area), mixed transfer (orange area), and unstable transfer (blue area) among the 25 sets of parameters. The short circuit and globular transfer modes exhibit distinctly representative behaviors of droplet transfer, as shown in Figure 3. Therefore, ultrasound is applied in short circuit transfer and globular transfer modes for further investigation. In the conventional GMAW process, the expansion and explosion of droplets are notably evident. In Figure 3a, the droplet expands and then explodes upon contact with the molten pool during short circuit transfer. In Figure 3b, the molten droplet expands dramatically and then explodes, turning into a smaller and incomplete shape.

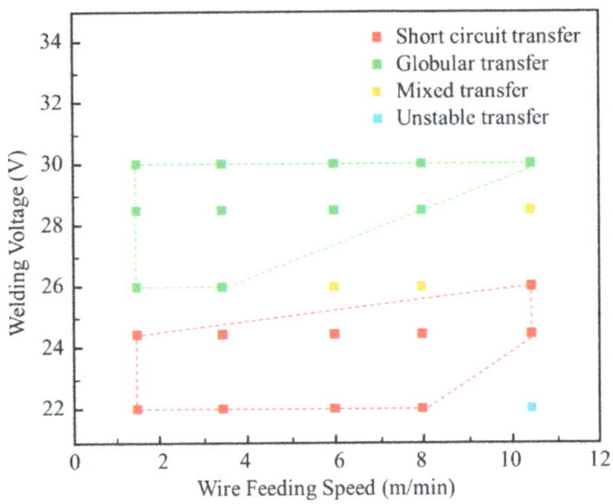

Figure 2. Parameter range of conventional GMAW droplet transfer modes.

(a)
(b)

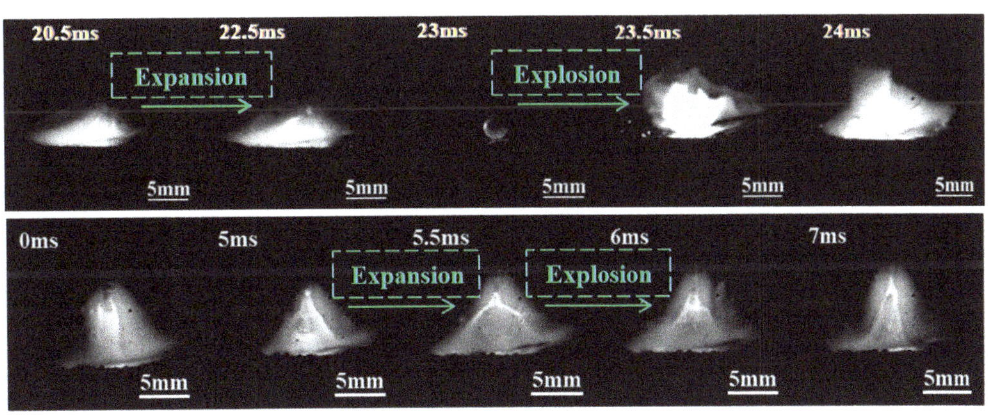

Figure 3. Droplet transfer process of short circuit transfer and globular transfer. (**a**) Short circuit transfer, U = 24.5 V, WFS = 6 mm/min; (**b**) Globular transfer, U = 28.5 V, WFS = 6 mm/min.

3.2. Effect of Ultrasound on Short Circuit Transfer Behavior

The ultrasonic output power is set to 0.8, 1.2, 1.6, and 2.0 kW to investigate the effect of ultrasonic power on the short circuit transfer. Welding voltage is 24 V, WFS is 6 m/min. Figure 4. Illustrates the droplet transfer behavior captured by the high-speed camera. The arc is significantly compressed compared to conventional GMAW in Figure 3a. The droplet still notably expands at the end of the wire, and a certain amount of nitrogen gas escapes, leading to nitrogen loss. Then the droplet continues to expand until it contacts the weld pool, forming a short circuit transfer. When the ultrasonic output power is 1.2 kW, the large-sized expansion of droplets decreases, and the surface shape of the droplets shows a wavy fluctuation due to the pressure difference between the inside and outside of the droplet. Local fragmentation and escape of a small amount of nitrogen gas occur in some areas occasionally, mainly due to insufficient surface tension and ultrasonic radiation force to balance the internal pressure exerted by nitrogen gas within the droplet. At other times, intense expansion can cause the rupture of molten droplets. When the ultrasonic output power increases to 1.6 kW and 2.0 kW, the droplet transfer phenomenon is similar to that at 1.2 kW, although the ultrasonic radiation force increases, there are still instances of droplet fragmentation and a certain degree of expansion and explosion, but the frequency Is relatively less. Increasing the ultrasonic power also Increases transfer frequency, causing the droplets to contact the weld pool before experiencing large-sized expansion, leading to reduced nitrogen escape.

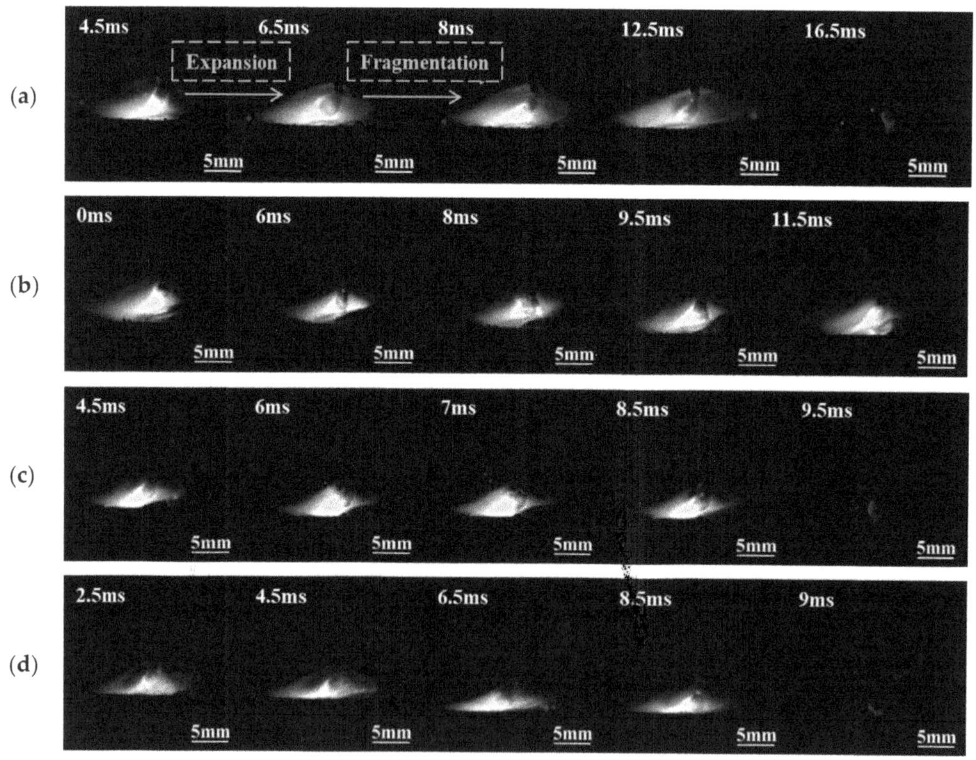

Figure 4. Influence of ultrasonic power on short circuit transfer. (**a**) 0.8 kW; (**b**) 1.2 kW; (**c**) 1.6 kW; (**d**) 2.0 kW.

The electric signals of the droplet transfer process were collected under conventional GMAW and different ultrasonic power levels, as shown in Figure 5. Under conventional GMAW conditions, the droplet transfer frequency is low, with a peak current reaching 400 A and a peak voltage of 40 V. When an ultrasonic power of 0.8 kW is applied, the number of wave peaks increases, and the peak current decreases to approximately 370 A, and the peak voltage decreases to around 37 V. As the ultrasonic power is increased to 1.2 kW and

1.6 kW, the number of wave peaks increases, and the peak current and voltage further decrease. At 2.0 kW ultrasonic power, compared to the previous conditions, the number of current and voltage wave peaks is the highest and the lowest, while the peak current drops to around 350 A and the peak voltage decreases to around 35 V. Analysis indicates that under conventional GMAW conditions, the re-ignition time of the arc after the droplet short circuit is long and the droplet transfer is unstable. However, with ultrasonic assistance, an increase in ultrasonic output power accelerates droplet transfer frequency, while the peak current and voltage decrease during short circuit transfer, indicating a shortened re-ignition time after the short circuit transfer and a more stable droplet transfer process.

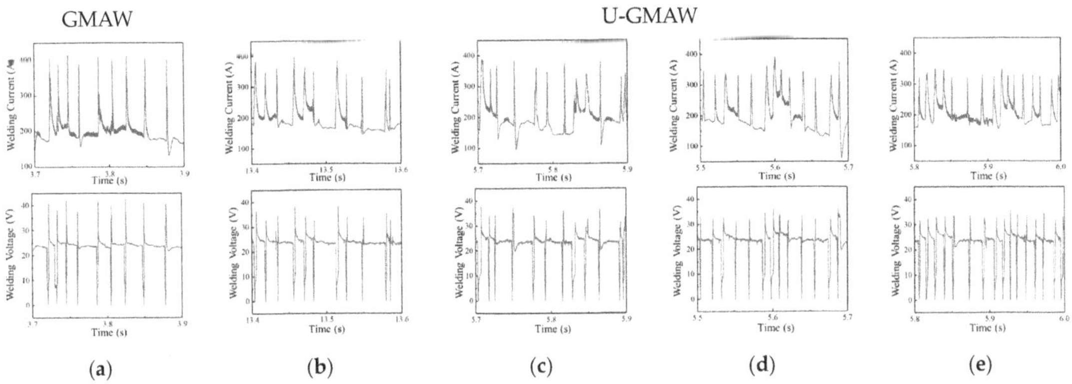

Figure 5. Current and voltage waveforms of conventional GMAW and U-GMAW. (**a**) Conventional GMAW; (**b**) 0.8 kW; (**c**) 1.2 kW; (**d**) 1.6 kW; (**e**) 2.0 kW.

Statistical analysis of droplet characteristic parameters at different ultrasonic power levels is shown in Figure 6. As the ultrasonic output power increases, the arc length gradually decreases, reaching a minimum at 2.0 kW. With the increase of ultrasonic power, the short circuit transfer frequency increases, and the droplet size decreases, exhibiting a reduction in numerical dispersion. This suggests that at higher ultrasonic power, the expansion of large droplets is suppressed, and the short circuit transfer frequency becomes more stable.

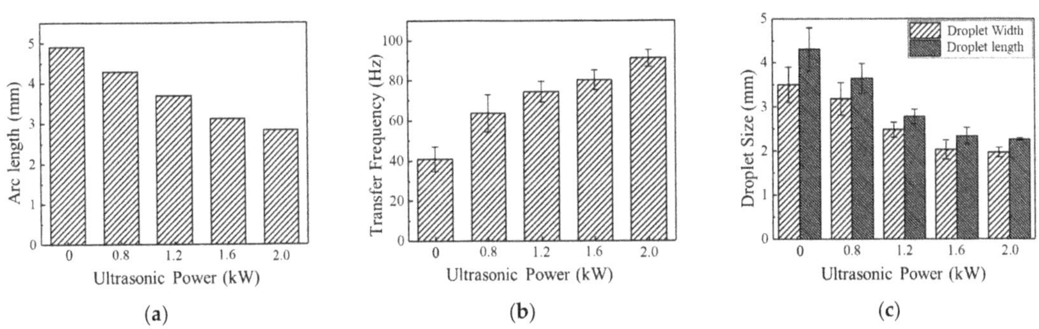

Figure 6. Short circuit transfer characteristic parameters under different ultrasonic powers. (**a**) Arc length; (**b**) Transfer frequency; (**c**) Droplet size.

3.3. *Effect of Ultrasound on Globular Transfer Behavior*

Figure 7 illustrates the globular transfer behavior under different ultrasonic power levels with a *WFS* of 6 m/min. At lower ultrasonic powers (0.8 kW and 1.2 kW), the arc length is significantly reduced compared to conventional GMAW in Figure 3b. However, droplets still expand and explode, suggesting insufficiency in ultrasonic radiation force. When the ultrasonic output power reaches 1.6 kW and 2.0 kW, the arc length decreases further, with a marked reduction in large-sized expansion. Nonetheless, some degree of expansion persists, and irregular surface shapes are noted in the droplets. Local fragmentation of droplets is observed, yet nitrogen emissions from fragmentation are lower than those resulting from large-scale bursting under conventional GMAW conditions. This indicates that ultrasonic assistance, to some extent, reduces nitrogen escape during droplet transfer in high nitrogen steel welding, thereby reducing nitrogen loss.

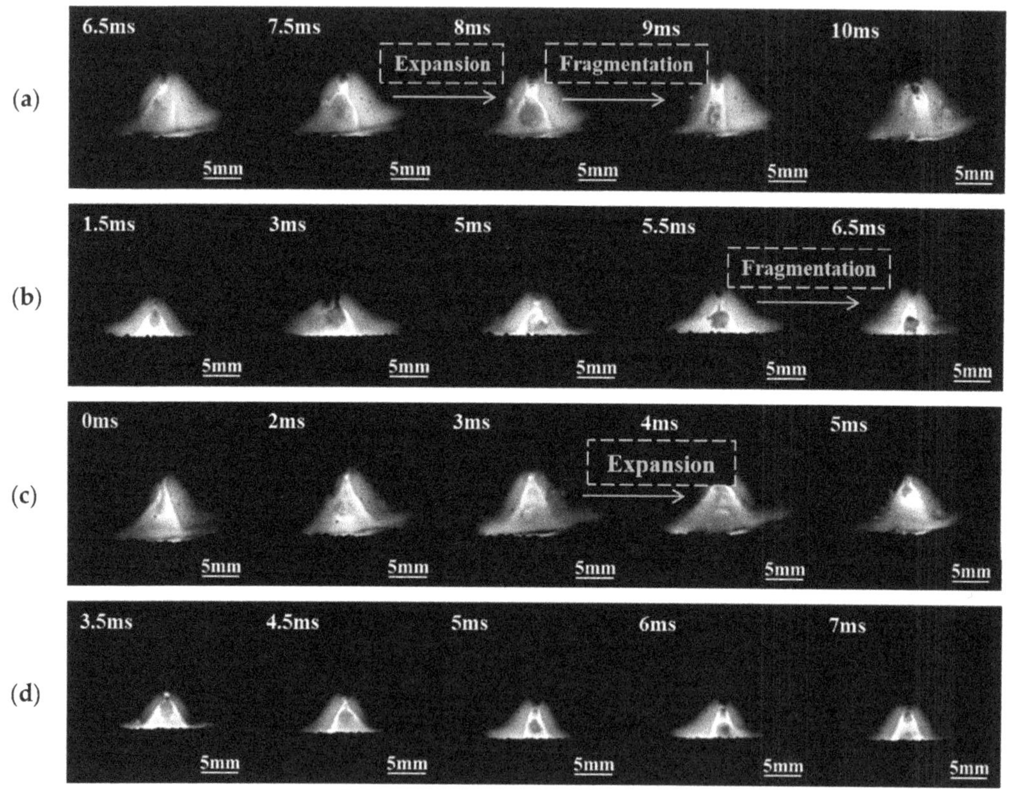

Figure 7. Influence of ultrasonic power on globular transfer. (**a**) 0.8 kW; (**b**) 1.2 kW; (**c**) 1.6 kW; (**d**) 2.0 kW.

The electrical signals at different ultrasonic power were collected, as shown in Figure 8. The current fluctuation range is 170 A to 270 A, and the welding voltage fluctuation range is 20 V to 40 V under conventional GMAW. The large fluctuations in the current and voltage values indicate the instability of the arc and the droplet transfer during globular transfer in GMAW. With the assistance of 0.8 kW ultrasound, the current fluctuation range narrows to 180 A to 250 A, indicating an improvement in the stability of the droplet transfer process. As the ultrasonic output power increases, the fluctuation range of the current and voltage values progressively decreases, thereby enhancing the stability of droplet transfer while mitigating the expansion and bursting of large droplets.

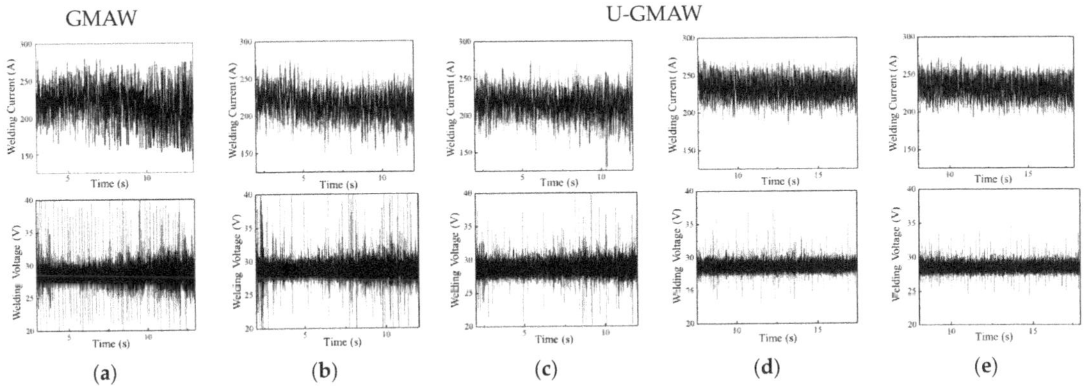

Figure 8. Current and voltage waveforms of conventional GMAW and U-GMAW. (**a**) conventional GMAW; (**b**) 0.8 kW; (**c**) 1.2 kW; (**d**) 1.6 kW; (**e**) 2.0 kW.

Figure 9 illustrates the droplet transfer characteristic parameters under different ultrasonic power levels. As ultrasonic power increases, the ultrasonic radiation force increases, causing the arc length to gradually decrease, accompanied by a reduction in droplet transfer period, bursting frequency, and droplet size, reaching the minimum values at an ultrasonic output power of 2.0 kW. This indicates the effective inhibition of large-scale droplet expansion under high ultrasonic power conditions. Regarding numerical dispersion, the magnitude of numerical variations decreases as ultrasonic output power increases. A sharp reduction in numerical dispersion is observed under ultrasonic power of 1.2 kW and 1.6 kW, where most droplets exhibit stable transfer with relatively minor size expansion. The most stable droplet transfer is achieved under 2.0 kW.

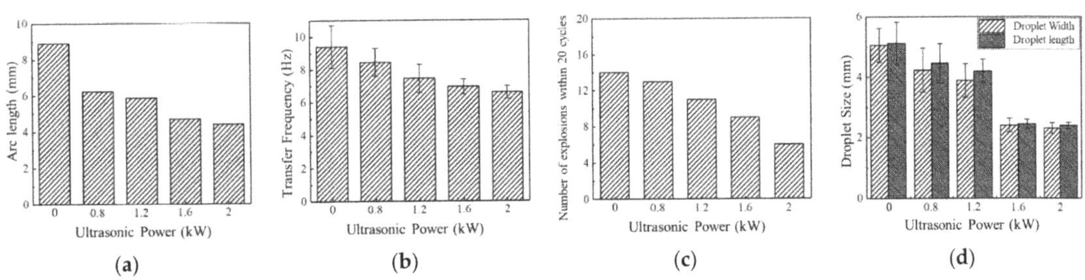

Figure 9. Globular transfer characteristic parameters at different ultrasonic powers. (**a**) Arc length; (**b**) Transfer frequency; (**c**) Number of explosions within 20 cycles; (**d**) Droplet size.

4. Discussion

Based on the arc constriction and droplet size reduction induced by ultrasonic radiation force, coaxial ultrasonic assistance presents a promising approach to controlling droplet transfer behavior in GMAW of high nitrogen steel. The discussion on the mechanism of droplet transfer in ultrasonic-assisted GMAW is as follows.

4.1. Mechanism of Ultrasound on Short Circuit Transfer

The process diagram of short circuit transfer behavior in GMAW and U-GMAW is presented in Figure 10. During the arcing phase, molten droplets form at the tip of the welding wire. As the droplets grow, a short circuit occurs when they contact with the weld pool. Under the combined effects of electromagnetic forces and surface tension, necking occurs, separating the molten droplet from the welding wire. The arc then re-ignites to initiate a new transfer cycle. While the short circuit transfer process in U-GMAW is similar to conventional GMAW, it features an increased transfer frequency and decreased arc length.

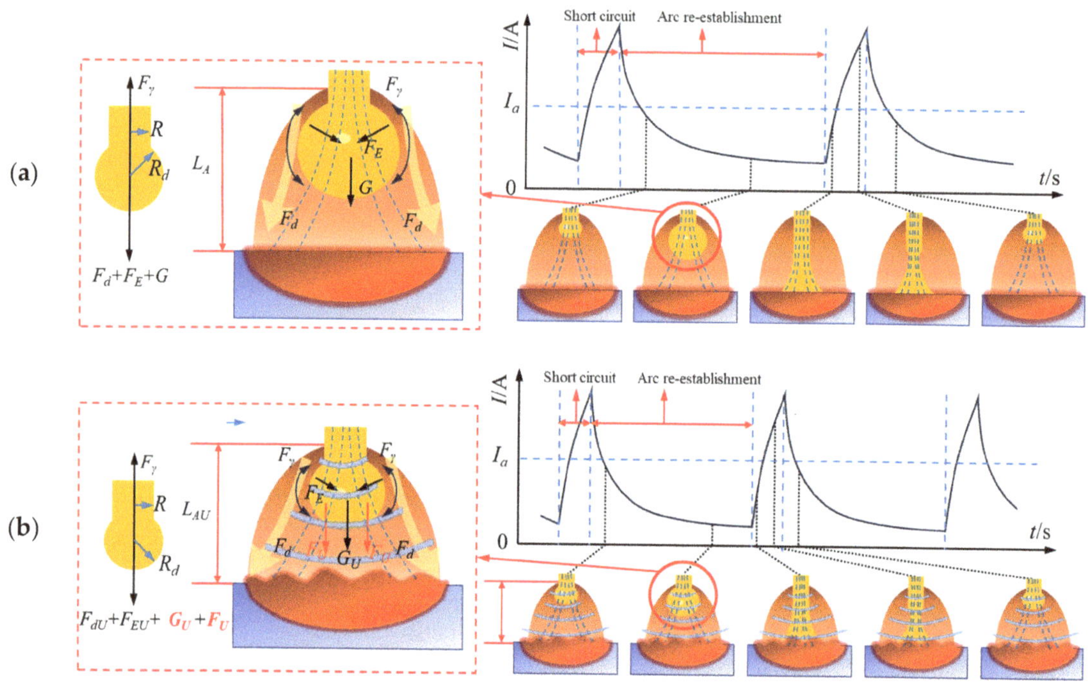

Figure 10. Schematic of the short circuit transfer process of GMAW and U-GMAW. (**a**) GMAW; (**b**) U-GMAW.

During conventional GMAW, as illustrated in Figure 10a, the molten droplets at the tip of the welding wire are subjected to electromagnetic force F_m, plasma flow force F_d, gravity G, and surface tension F_γ. The combined forces of F_m, F_d, and G act downward, promoting droplet transfer. When the droplet has not yet detached from the welding wire, F_γ prevents the droplet from falling, as expressed by Equation (1) (where R represents welding wire diameter and γ represents surface tension coefficient). Notably, F_γ is not significantly influenced by ultrasound.

$$F_\gamma = 2\pi R\gamma \quad (1)$$

Equation (2) is the expression for the electromagnetic force F_m acting on a unit volume, where $\vec{J}$ represents the current density, and $\vec{B}$ represents the magnetic field line density.

$$F_m = \vec{J} \times \vec{B} \quad (2)$$

During the welding process, droplets serve as channels for current flow and are also subjected to electromagnetic forces. According to the high-speed camera results, under the experimental conditions of this study, the arc nearly envelops the molten droplets, causing F_m to promote the downward transfer of the molten droplets [19], as expressed by Equation (3).

$$F_m = \frac{\mu_0 I^2}{4\pi}\left(\text{In}\frac{R_d \sin\theta}{R} - \frac{1}{4} - \frac{1}{1-\cos\theta} + \frac{2}{(1-\cos\theta)^2}\text{In}\frac{2}{1-\cos\theta}\right) \tag{3}$$

where μ_0 represents the vacuum permeability, I represents the welding current, R_d represents the droplet diameter, and θ is the angle between the line connecting the droplet center and the contact point of the arc on the droplet and the direction of gravity. In this study, the molten droplets are enveloped by the arc, making θ greater than 60° degrees. The coefficient in parentheses of Equation (3) is close to 1, and the electromagnetic force can be expressed in Equation (4). The electromagnetic force in this study is mainly affected by welding current.

$$F_m = \frac{\mu_0 I^2}{4\pi} \tag{4}$$

Plasma flow force F_d, detailed in Equation (5), also has an effect on molten droplets [20]:

$$F_d = 0.5 C_d \rho_g v_g^2 \pi R_d^2 \left(1 - \frac{R^2}{2R_d^2}\right) \tag{5}$$

where C_d represents the plasma flow coefficient, related to the droplet diameter D_d and viscosity μ, ρ_g represents the gas density, and v_g represents the gas flow rate.

The droplet gravity is expressed in Equation (6):

$$G = \rho V g \tag{6}$$

where ρ represents the density of molten droplets, V represents droplet volume, and g represents gravitational acceleration.

During U-GMAW, the ultrasonic device emits ultrasound in a downward direction, generating a standing wave sound field and exerting acoustic radiation pressure on the molten droplets. The ultrasonic radiation force F_U is expressed in Equation (7), where ρ_m represents the density of the sound propagation medium, A represents the velocity amplitude of the incident wave, k represents the number of waves, R_d represents the droplet radius m, and λ_p represents the ratio of medium density to droplet density [18].

$$F_U = \frac{1}{3}\pi\rho_m |A|^2 (kR_d)^3 \sin(2kd)\frac{5-2\lambda_p}{2+\lambda_p} \tag{7}$$

Based on the research results mentioned above, it is evident that the introduction of ultrasound compresses the molten droplets, reducing their volume and consequently lowering the gravitational force ($G_U < G$), which impedes droplets from making contact with the weld pool below. Reduced droplet size also leads to a decrease in F_{dU} ($F_{dU} < F_d$). The electromagnetic force will slightly decrease ($F_{mU} < F_m$), as ultrasound will cause a decrease in the peak current of short circuit transfer. However, the downward pressure exerted by the ultrasound field (F_U) compensates for the reduction in G_U, F_{dU}, and F_{mU}. The force exerted by ultrasound on molten droplets is the main reason for promoting droplet transfer.

The ultrasound significantly compresses the arc, reducing its length and facilitating easier contact between the molten droplets and the weld pool, thereby promoting short circuit transfer. Additionally, ultrasound radiation forces oscillate the molten weld pool, causing surface fluctuations that facilitate faster contact between droplets and the weld pool, thereby accelerating the transfer frequency. Consequently, ultrasound promotes short circuit transfer, allowing the droplets to make contact with the weld pool and undergo short circuit transfer even when the nitrogen bubble aggregation in the droplets is minimal, thus reducing nitrogen loss.

4.2. Mechanism of Ultrasound on Globular Transfer

The schematic depiction of the globular transfer behavior of GMAW and U-GMAW is presented in Figure 11. The droplet transfer process differs from that of short circuit transfer, where the molten droplet detaches from the welding wire tip and then enters the weld pool. Similar to short circuit transfer, ultrasound radiation force compresses the droplet, reducing nitrogen accumulation.

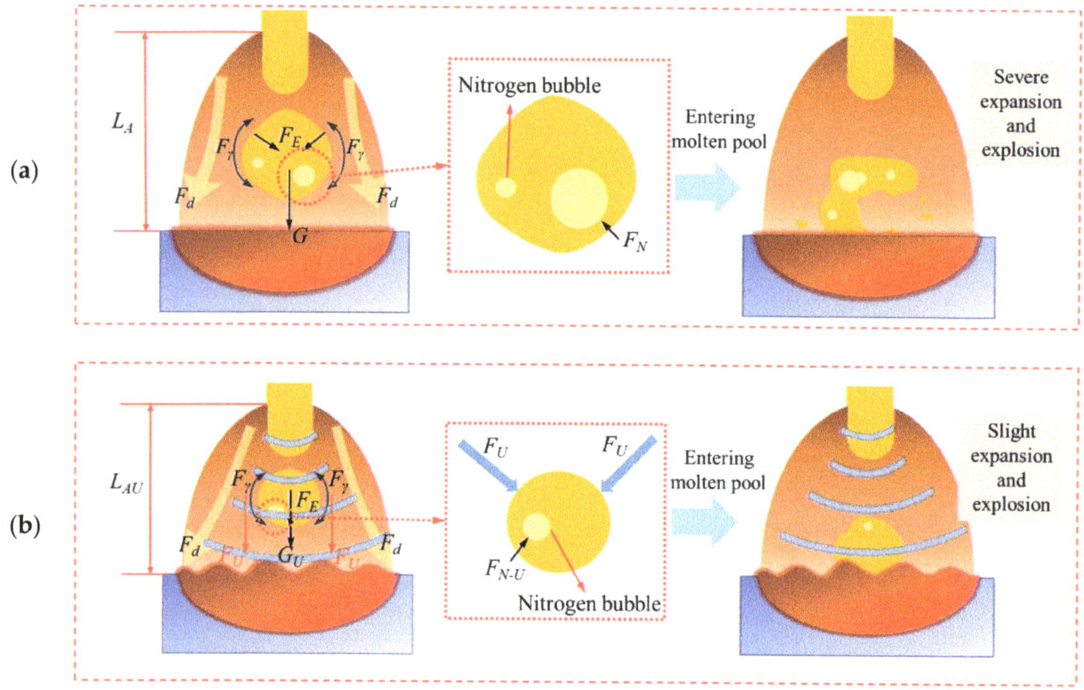

Figure 11. Schematic of the globular transfer process of GMAW and U-GMAW. (**a**) GMAW; (**b**) U-GMAW.

Additionally, ultrasound has an inhibitory effect on the formation of nitrogen bubbles. Research has shown that the factors affecting the solubility of nitrogen in steel mainly include nitrogen partial pressure, temperature, and alloy composition [21]. During the droplet transfer process, the alloy composition of high nitrogen steel droplets remains unchanged. Given the rapid droplet transfer process, and assuming that the temperature of the droplets remains constant from the welding wire tip to entering the molten pool, nitrogen partial pressure is the primary factor affecting the solubility of nitrogen in molten droplets. Nitrogen dissolved in molten iron exists in the form of nitrogen ions, but nitrogen in molten steel often exists in the form of nitrogen atoms [22]. The reaction of nitrogen dissolution in liquid steel is:

$$\frac{1}{2}N_2(g) = [N] \tag{8}$$

K_N is the equilibrium constant of nitrogen dissolution reaction, as measured in Equation (9) [23], where a_N and f_N are the nitrogen activity and Henrian activity coefficients of the assumed 1% N alloy melt as reference states, [% N] represents the mass percentage of dissolved nitrogen, P_{N_2} is the partial pressure of nitrogen, and P^0 is a standard atmospheric pressure of 1.01325×10^5 Pa.

$$K_N = \frac{a_N}{\sqrt{P_{N_2}/P^0}} = \frac{f_N[\%N]}{\sqrt{P_{N_2}/P^0}} \tag{9}$$

Taking the logarithm of both sides of Equation (9) yields Equation (10):

$$\lg[\%N] = \frac{1}{2}\lg\left(P_{N_2}/P^0\right) + \lg K_N - \lg f_N \tag{10}$$

The K_N of steel can be expressed as Equation (11) [24], and K_N can be considered a constant at a constant temperature.

$$\lg K_N = -\frac{188}{T} - 1.17 \tag{11}$$

f_N is mainly affected by temperature and alloy composition [25]. During the droplet transfer process, the composition and temperature changes of the droplets are negligible, and f_N is considered a constant value. Thus, the solubility of nitrogen is mainly influenced by the partial pressure of nitrogen. The dissolved nitrogen concentration in the droplet during the droplet transfer of GMAW and U-GMAW is primarily determined by the nitrogen partial pressure. According to Equation (11), [% N] and $(P_{N_2}/P^0)^{1/2}$ are approximately proportional, meaning the solubility of nitrogen increases as the nitrogen partial pressure increases. The ultrasonic radiation applied to droplets increases the pressure ($P_{N_2\text{-}U} > P_{N_2}$) on nitrogen bubbles in the droplets, resulting in an increase in nitrogen solubility. This effect mitigates the formation of nitrogen bubbles and subsequently reduces the tendency for nitrogen escape. Furthermore, as ultrasonic power intensifies, the radiation force exerted by ultrasound increases, further reducing the propensity of nitrogen escape.

5. Conclusions

The study investigated the droplet transfer behavior of wires with nitrogen contents of 0.35% in ultrasonic-assisted GMAW (U-GMAW), focusing on short circuit transfer and globular transfer. The influence of ultrasonic output power on droplet transfer behavior was investigated, and the underlying mechanisms were analyzed. The main conclusions are summarized as follows:

1. Ultrasonic power has a significant influence on the droplet transfer process of HNS welding wire. Ultrasonic assistance improves the stability of droplet transfer: as ultrasonic power increases, the arc is further compressed, leading to a reduction in arc length. Droplet expansion is effectively inhibited under both short circuit transfer and globular transfer modes.
2. The influence of ultrasonic assistance on welding electrical signals was studied. It was found that the droplet transfer frequency is lower during conventional GMAW short circuit transfer, with relatively high peak current and voltage under both transfer modes. As the ultrasonic power increases, the short circuit transfer frequency increases, and the peak current and voltage under both modes decrease, making the droplet transfer process more stable.
3. The behavior mechanisms of short circuit transfer and globular transfer in U-GMAW were analyzed. The ultrasonic radiation force compresses the arc length, accelerates droplet transfer frequency, and suppresses droplet expansion, thereby reducing nitrogen accumulation within the droplets. Consequently, more nitrogen elements can be transferred into the molten pool during welding with HNS welding wires, leading to a reduction in nitrogen loss.

Author Contributions: Conceptualization, C.F.; formal analysis, J.L., Z.H. (Zhizheng He), Z.L., Z.H. (Zihuan Hua) and B.T.; investigation, Z.H. (Zihuan Hua); writing—original draft preparation, J.L.; writing—review and editing, J.L., Z.H. (Zhizheng He), Z.L., B.T. and C.F.; visualization, J.L.; supervision, C.F.; project administration, C.F.; funding acquisition, C.F. All authors have read and agreed to the published version of the manuscript.

Funding: This research received no external funding.

Institutional Review Board Statement: Not applicable.

Informed Consent Statement: Not applicable.

Data Availability Statement: The raw data supporting the conclusions of this article will be made available by the authors on request.

Conflicts of Interest: The authors declare no conflicts of interest.

References

1. Simmons, J.W. Overview: High-nitrogen alloying of stainless steels. *Mater. Sci. Eng. A* **1996**, *207*, 159–169. [CrossRef]
2. Berns, H. Manufacture and application of high nitrogen steels. *Int. J. Mater. Res.* **1995**, *86*, 156–163. [CrossRef]
3. Lang, Y.-P.; Qu, H.-P.; Chen, H.-T.; Weng, Y.-Q. Research progress and development tendency of nitrogen-alloyed austenitic stainless steels. *J. Iron Steel Res. Int.* **2015**, *22*, 91–98. [CrossRef]
4. Yang, K.; Ren, Y. Nickel-free austenitic stainless steels for medical applications. *Sci. Technol. Adv. Mater.* **2010**, *11*, 014105. [CrossRef] [PubMed]
5. Sumita, M.; Hanawa, T.; Teoh, S.H. Development of nitrogen-containing nickel-free austenitic stainless steels for metallic biomaterials. *Mater. Sci. Eng. C* **2004**, *24*, 753–760. [CrossRef]
6. Talha, M.; Behera, C.K.; Sinha, O.P. A review on nickel-free nitrogen containing austenitic stainless steels for biomedical applications. *Mater. Sci. Eng. C* **2013**, *33*, 3563–3575. [CrossRef]
7. He, S.; Yang, D.; Huang, Y.; Wang, K. Effect of the current waveform on the droplet transfer in CMT welding high-nitrogen steel. *J. Manuf. Process.* **2022**, *75*, 41–48. [CrossRef]
8. Makaya, A.; Fredriksson, H. Study on the production of Fe–Cr–Mn–C–Si foam by nitrogen solubility difference between the liquid and solid phases. *Mater. Sci. Eng. A* **2005**, *413*, 533–537. [CrossRef]
9. Wang, L.; Li, Y.; Ding, J.; Xie, Q.; Zhang, X.; Wang, K. Problems in welding of high nitrogen steel: A review. *Metals* **2022**, *12*, 1273. [CrossRef]
10. Dimitrov, V.I.; Jekov, K.; Avinc, A. Prediction of the solubility of nitrogen in steels obtained by pressurised electroslag remelting process. *Comput. Mater. Sci.* **1999**, *15*, 400–410. [CrossRef]
11. Li, J.; Li, H.; Liang, Y.; Liu, P.; Yang, L. The microstructure and mechanical properties of multi-strand, composite welding-wire welded joints of high nitrogen austenitic stainless steel. *Materials* **2019**, *12*, 2944. [CrossRef] [PubMed]
12. Hosseini, V.A.; Wessman, S.; Hurtig, K.; Karlsson, L. Nitrogen loss and effects on microstructure in multipass TIG welding of a super duplex stainless steel. *Mater. Des.* **2016**, *98*, 88–97. [CrossRef]
13. Yang, D.; Xiong, H.; Huang, Y.; Yan, D.; Li, D.; Peng, Y.; Wang, K. Droplet transfer behavior and weld formation of gas metal arc welding for high Nitrogen austenitic stainless steel. *J. Manuf. Process.* **2021**, *65*, 491–501. [CrossRef]
14. Liu, Z.; Fan, C.; Chen, C.; Ming, Z.; Yang, C.; Lin, S.; Wang, L. Design and evaluation of nitrogen-rich welding wires for high Nitrogen stainless steel. *J. Mater. Process. Technol.* **2021**, *288*, 116885. [CrossRef]
15. Yang, D.; Fang, H.; Peng, Y.; Fan, J.; Huang, Y.; Wang, K.; Yan, D.; Li, D. Investigation of spatters in cold metal transfer+ pulse-based wire and arc additive manufacturing of high nitrogen austenitic stainless steel. *J. Mater. Eng. Perform.* **2021**, *30*, 6881–6894. [CrossRef]
16. Cui, B.; Zhang, H.; Liu, F.D. Effects of shielding gas composition on the welding stability, microstructure and mechanical properties in laser-arc hybrid welding of high nitrogen steel. *Mater. Res. Express* **2018**, *5*, 096513. [CrossRef]
17. Cui, B.; Chen, K.; Yang, Y.; Lv, Y.; Zhang, F.; Liu, S. Effect of ultrasonic vibration on the pores and properties of the laser-arc hybrid welding joint of high nitrogen steel. *Mater. Chem. Phys.* **2024**, *318*, 129297. [CrossRef]
18. Society, A.W.; Chen, C.; Fan, C.; Cai, X.; Liu, Z.; Lin, S.; Yang, C. Characteristics of arc and metal transfer in pulsed ultrasonic-assisted GMAW. *Weld. J.* **2020**, *99*, 203s–208s. [CrossRef]
19. Zhang, S.; Ma, G.; Peng, X.; Xiang, Y. Numerical simulation of the effects of bypass current on droplet transfer during AZ31B magnesium alloy DE-GMAW process based on FLUENT. *Int. J. Adv. Manuf. Technol.* **2017**, *90*, 857–863. [CrossRef]
20. Moelans, N.; Blanpain, B.; Wollants, P. An introduction to phase-field modeling of microstructure evolution. *Calphad* **2008**, *32*, 268–294. [CrossRef]
21. Sano, M.; Fujita, Y.; Mori, K. Formation of bubbles at single nonwetted nozzles in mercury. *Metall. Trans. B* **1976**, *7*, 300–301. [CrossRef]
22. Ishii, F.; Ban-Ya, S.; Fuwa, T. Solubility of nitrogen in liquid iron and iron alloys containing the group VIa elements. *Tetsu-to-Hagané* **1982**, *68*, 946–955. [CrossRef] [PubMed]

23. Anson, D.; Pomfret, R.J.; Hendry, A. Prediction of the solubility of nitrogen in molten duplex stainless steel. *ISIJ Int.* **1996**, *36*, 750–758. [CrossRef]
24. Rawers, J.C.; Gokcen, N.A. High-temperature, high-pressure nitrogen concentration in Fe-Cr-Mn-Ni alloys. *Steel Res.* **1993**, *64*, 110–113. [CrossRef]
25. Satir-Kolorz, A.H.; Feichtinger, H.K. On the solubility of nitrogen in liquid iron and steel alloys using elevated pressure. *Int. J. Mater. Res.* **1991**, *82*, 689–697. [CrossRef]

materials

MDPI

Article

Laser Welding of ARMOX 500T Steel

Aleksander Lisiecki [1,*], Agnieszka Kurc-Lisiecka [2], Wojciech Pakieła [3], Grzegorz Chrobak [4], Gilmar Ferreira Batalha [5] and Marcin Adamiak [6]

[1] Department of Welding Engineering, Faculty of Mechanical Engineering, Silesian University of Technology, Konarskiego 18A Str., 44-100 Gliwice, Poland
[2] Institute of Applied Sciences, WSB Merito University in Poznan, Sportowa 29 Str., 41-506 Chorzow, Poland; agnieszka.kurc-lisiecka@chorzow.merito.pl
[3] Department of Engineering Materials and Biomaterials, Silesian University of Technology, Konarskiego 18A Str., 44-100 Gliwice, Poland; wojciech.pakiela@polsl.pl
[4] IPG Photonics Sp. z o.o., ul. Wyczółkowskiego 8 Str., 44-109 Gliwice, Poland; gchrobak@ipgphotonics.com
[5] Department of Mechatronics and Mechanical Systems Engineering, Polytechnic School of Engineering of the University of Sao Paulo (USP), São Paulo 05508-900, Brazil; gfbatalh@usp.br
[6] Materials Research Laboratory, Faculty of Mechanical Engineering, Silesian University of Technology, Konarskiego 18A Str., 44-100 Gliwice, Poland; marcin.adamiak@polsl.pl
* Correspondence: aleksander.lisiecki@polsl.pl

Abstract: The article describes the results of the study on laser welding of armor plates with a nominal thickness of 3.0 mm. The plates were made of Armox 500T steel characterized by a hardness of up to 540 HB, a minimum yield strength of 1250 MPa, an ultimate strength of up to 1750 MPa, and an elongation A5 minimum of 8%. The laser used for the welding tests was a solid state Yb:YAG laser. The influence of basic parameters such as laser output power, welding speed, and focal plane position on the weld geometry was determined during bead-on-plate welding tests. The optimal conditions for butt joint welding were determined, and the test joints were subjected to mechanical and impact tests, metallographic analysis, and hardness measurements. It has been shown that it is possible to laser weld Armox 500T armor plates, and at the same time it is possible to provide high quality butt joints, but this requires precise selection of welding parameters. A decrease in HAZ hardness of about 22–35% in relation to the hardness of the base material, ranging from 470 to 510 HV0.2, was found. The ultimate tensile strength of the test joints was approx. 20% lower than the Armox 500T steel. The bending tests revealed the low plasticity of the tested joints because the bending angle was just 25–35°. The results of Charpy V–notch test revealed that the impact toughness of the weld metal at −20 °C was approx. 30% lower than at room temperature.

Keywords: laser welding; armor steel; Armox 500T; microstructure; mechanical properties

1. Introduction

Armored vehicles must provide high ballistic protection, and on the other hand, they must have the lowest possible weight to ensure high mobility even in very difficult terrain and environments [1]. The basic structural material for such vehicles is steel because it is susceptible to plastic processing, joined via welding, and facilitates the shaping of functional properties in a wide range as a result of heat treatment [1,2]. Steels that provide good ballistic properties are usually ultra-high strength (UHS) steels processed by quenching and tempering (Q&T). High mechanical properties are provided by martensitic microstructure. However, due to subsequent tempering, the ductility and toughness is increased, providing better impact energy absorption which is crucial for ballistic performance. Many researchers involved in testing the ballistic characteristic of armor plates made of different grades of armor steel indicate that the higher the hardness of the steel obtained as a result of tempering, the higher the ballistic resistance [3–5]. Therefore, tempering takes place at a relatively low temperature, retaining high hardness and thus high ballistic performance [1,6].

The most commonly used joining methods of the armor steel components are gas metal arc welding (GMA) and submerged arc welding (SAW). Recent research has also been conducted on hybrid laser arc welding (HLAW) [1,7–10]. Fei et al. [1] presented a novel technique of TIG welding of armor steel with a trapezoidal AISI309 austenitic stainless steel interlayer. Thanks to controlling the dilution and sequence of solidification, they provided the desired chemical composition and microstructure of the weld metal, contributing improvements in strength and toughness, as well as limiting the tendency toward both solidification cracking and hydrogen-assisted cold cracking.

Turichin et al. [7] investigated laser arc hybrid welding of ultra-high strength steels, including Armox 600T steel. They used three filler wires of different chemical compositions and determined the influence of the filler wires' chemical composition on the microstructure and mechanical properties of the welds. The main conclusion drawn from the investigations was that the welds of 7.0 mm thick Armox 600T plates were about 14% weaker than the base metal.

Kumar et al. [8] conducted a study on shielded metal arc welding of ultra-high hard armor steel of CE equivalent 0.91 using the following different electrodes: austenitic stainless steel type, super duplex stainless steel type, and low-hydrogen ferritic type. They found that the joints made via the austenitic stainless steel electrodes showed the best impact toughness, while the joints made via low-hydrogen ferritic electrodes exhibited the best tensile strength and the highest hardness.

Skowrońska et al. [9] also investigated a hybrid process of armor steel welding. They investigated PTA-MAG (plasma transferred arc—metal active gas) welding of RAMOR 500 steel plates 6.7 mm thick with a filler type Mn4Ni2CrMo. They pointed some advantages of the welding process, including limited decreasing of mechanical properties of HAZ at an acceptable level from the point of view of maintaining the antiballistic properties of the base material.

All of the mentioned methods usually utilize austenitic type wire as the additional material because the weldability of the armor grade steel is limited due its high carbon equivalent, high hardness, and thus tendency for cold cracking. However, the use of a plastic austenitic deposit leads to a significant reduction in the ballistic performance of welded joints, as reported by Fei et al. [1] and Magudeeswaran et al. [6]. Moreover, even the plastic and relatively soft austenitic deposit does not completely solve the problem of hydrogen-induced cracking (HIC) in the heat-affected zone (HAZ) of the welded joints of armor steel [6].

A method that has not yet been used industrially for the welding of armor steel, but provides some benefits and gives the possibility of further optimization and development of welding technology of this grade of steel is laser welding [11–17]. Laser welding may be conducted as an autogenous process, without additional material, and thanks to efficient shielding of inert gas, the process can be considered as a low-hydrogen process. Moreover, the characteristic feature of laser welding is a narrow HAZ and a relatively low level of welding stresses, which in turn is beneficial to the reduction in the tendency for hydrogen-induced cracking. It is also worth noting that currently available lasers with a power of several dozen kW allow for welding joints with a thickness over 30 mm in one pass. The studies carried out so far in the field of laser welding of armor steel indicate that despite the low heat input of laser welding and rapid cooling, softening of HAZ occurs, which may be additionally beneficial in terms of the reduction in the tendency for hydrogen-induced cracking. Janicki [11] investigated the autogenous laser welding of butt joints of 3.6 mm thick Armox 500T steel, and within the investigated technological conditions of welding he reported a significant drop of hardness in the HAZ, up to 40% comparing to the base material. Similar results were reported by Bassett [12] who found 30% hardness drop in HAZ of laser welded 13.0 mm thick armor steel CP50.

Laser welding also shows some limitations. Since the laser beam is very a precise heat source, usually focused to a diameter of 0.2–0.3 mm, it requires high tolerance of edges preparation and positioning of the laser beam along the weld trajectory, in the case of butt joint welding. Moreover, in the case of autogenous welding, it is not possible to control the chemical composition of the weld metal, and weld undercuts are typical. These limitations make the application of laser welding on an industrial scale difficult in certain areas [15–24].

However, there are few publications in the world on the laser welding of armor steel. Therefore, research was undertaken on the technology of autogenous laser welding of armor plates in order to determine the influence of basic welding parameters on the microstructure and mechanical properties, in particular the plasticity and impact toughness of test joints also at low temperature. As a result, laser welding conditions in which weld porosity has been eliminated were determined. The obtained narrow welds, compared to arc and hybrid welding methods, may be beneficial for ensuring better ballistic performance of the entire butt joints of armor plates. Therefore, it is planned to continue research in this area, expanding the range of plates thicknesses and grades of armor steels.

Moreover, the preliminary tests of welding these armor plates with a single-mode beam and also via a unique technique with two combined laser beams, AMB (adjustable mode beam) with a total power of 6.0 kW, and beam oscillation have been already carried out. Results have shown that it is possible to further optimize the laser welding process, which ensures a favorable shape of the weld and favorable joint properties, even in the case of larger sheet thicknesses.

2. Materials and Methods

The armor plates with the dimensions of 250 × 250 mm used in this study were made of armor steel Armox 500T (SSAB, Stockholm, Sweden) in quenched and tempered condition (Q&T). The smallest thickness plate available from the manufacturer, intended for ballistic protection of light vehicles, was selected for the tests. The nominal thickness of the plates was 3.0 mm. Such an armor plate should provide protection against normal 7.62 mm caliber bullets (excluding armor-piercing bullets, e.g., with a tungsten core). Chemical composition and mechanical properties of Armox 500T steel, provided by manufacturer SSAB and determined within the study, are listed in Tables 1 and 2, respectively.

Table 1. Chemical composition of Armox 500T steel (wt%).

Accor. to	C	Mn	Si	P	S	Cr	Ni	Mo	B
SSAB	0.32	1.2	0.4	0.01	0.003	1.0	1.8	0.7	0.005
Tested	0.28	1.1	0.2	-	-	0.88	1.4	0.6	-

Remarks: SSAB: results provided by the manufacturer, tested: GDS analysis of 3.0 mm thick Armox 500T plate.

Table 2. Mechanical properties of Armox 500T steel (wt%).

Accor. to	Hardness, HBW	Minimum Yield Strength $Rp_{0.2}$, MPa	Tensile Strength Rm, MPa	Total Elongation, A_5, %	Total Elongation, A_{50}, %	Notch Impact Energy at −40 °C, KV, J
SSAB	480–540	1250	1450–1750	8	10	32
Tested	-	1298	1553	8.2	11	22

Remarks: SSAB: results provided by the manufacturer, tested: results determined during mechanical testing of 3.0 mm thick Armox 500T plate.

The laser welding tests were conducted on a fully automated stand with the solid state Yb:YAG Disk laser (Trumpf, Ditzingen, Germany) emitted in continuous wave (cw) mode at 1.03 μm wavelength with maximum output power of 3.3 kW, and beam quality (beam parameter product—BPP) of 8.0 mm·mrad. The laser beam was focused to a diameter of 200 μm.

In the initial stage of the study, the bead-on-plate welds were produced to simulate the process of autogenous butt joint welding and to investigate the influence of the basic welding parameters on the penetration depth, weld shape, and quality of welds (e.g., presence of voids, lack of penetration, undercuts, etc.). The initial tests were conducted in the laser output power range from 1.0 to 3.0 kW, welding speed from 0.5 to 2.5 m/min, and different focal position. The detailed bead-on-plate laser welding conditions are summarized in Table 3.

Table 3. Parameters of bead-on-plate laser welding of the 3.0 mm thick plates of Armox 500T steel.

Weld Bead No.	Output Laser Power kW	Welding Speed m/min	Focal Position * mm	Energy Input ** J/mm	Width of the Weld Face mm	Width of the Weld Root/Penetration Depth mm	Fusion Zone Shape	Remarks
B1	1.0	0.5	0	120	2.0	-/2.7	Y	L, P
B2	1.25	0.5	0	150	2.4	1.3	X	L, P
B3	1.25	0.5	1.5	150	3.0	0.3	X	F
B4	1.25	0.5	−1.5	150	2.5	1.3	Y	F, P
B5	1.25	0.5	−3.0	150	2.7	-/2.5	Y	L, P
B6	1.5	0.5	0	180	2.6	1.8	X	F
B7	1.5	1.0	0	90	2.7	1.2	Y	F
B8	3.0	2.0	0	90	1.8	2.5	X	F
B9	3.0	2.5	0	70	1.4	2.3	X	F,H

Remarks: * position relative to the top surface of plate (negative value means focusing below the surface); ** energy input is simply calculated by considering laser output power and welding speed only, while the heat input also takes into account the heat transfer coefficient; Quality assessment: L—lack of fusion, F—full penetration, H—hollow face, P—porosity.

Based on the initial tests, parameters providing full penetration and required weld quality (proper shape and no voids in weld) were chosen for autogenous butt joint laser welding. In order to determine the influence of energy input on the microstructure and mechanical properties of test joints, the extreme parameters with the highest (180 J/mm) and lowest (70 J/mm) energy input were chosen for laser welding of butt joints.

Specimens for welding tests were cut from the armor plate into pieces in dimensions of 120.0 × 45.0 mm via mechanical cutting. Prior to the welding test, the specimens were sandblasted and additionally the edges to be welded were cleaned with acetone.

The specimens to be welded were mounted in a stiffening clamping device with backing gas flow (argon of purity 99.999%) to protect the root side of the weld against ambient air. The argon flow from the root side was kept constant at 5.0 L/min. The top side of the welding zone was protected by argon flow via four cylindrical nozzles 8.0 mm in diameter each, and set at an angle of 45° to the sample surface. The flow of argon was kept at approx. 20.0 L/min. The experimental setup and a view of laser welding is shown in Figure 1.

When the laser welding tests were completed, first the visual inspections (VT) were performed according to PN-EN ISO 17637:2017-02 standard [25]. Next the metallographic examinations were carried out. The samples for metallographic examinations were cut perpendicularly to the weld axis away from the region of initiation and completion of welding.

The samples were mounted in thermosetting phenol resin with graphite filler Electro-WEM (Metalogis, Warsaw, Poland), and next the samples were wet ground via water papers with grit 120 to 2500 using an automatic grinding/polishing machine Struers Labopol-2 (Struers, Rodovre, Denmark). Next the cross-sections were polished with 1 μm diamond suspension Metkon Diapat-M (Metkon Instruments Inc., Bursa, Turkey). After polishing, the cross-sections were etched via Nital solution (3.0%).

Macrostructure analyses were carried out on the OLYMPUS SZX9 microscope (Olympus Corporation, Tokyo, Japan), while the microstructure observations were performed with a NIKON Eclipse MA100 microscope (Nikon Corporation, Tokyo, Japan). The microstructure was also examined via scanning electron microscopy (SEM) (Carl Zeiss, Oberkochen, Germany), equipped with the energy dispersive spectrometer (EDS) (Oxford Instruments, Abingdon, UK). The phase composition was determined via X-ray diffraction (Panalitycal, Almelo, The Netherlands) with CuKα source of radiation andwith the scanning range of the diffraction angle 2θ from 0 to 140°. Chemical composition of the investigated steel was determined via a glow discharge spectrometer GDS850 (LECO, Geleen, The Netherlands).

Figure 1. The autogenous laser welding of butt joints of Armox 500T steel samples.

The hardness distribution was measured on the cross-section of the test butt joints via the hardness tester WILSON WOLPERT 401 MVD (Wolpert Wilson Instruments, Aachen, Germany) at the load 5 N and the dwell time 10 s. The measurements were conducted in the middle of the plate thickness according to the PN-EN ISO 9015-2:2016-04 standard [26].

Mechanical properties of test joints were determined via the technological bending test (PN-EN ISO 5173:2023-06 [27], static tensile test (PN-EN ISO 148-1:2017-02 [28]), and Charpy V–notch test (PN-EN ISO 4136:2022-12 [29]).

The typical size of samples for impact tests according to the standard is 10 × 10 mm, but the standard allows samples with a smaller thickness, even up to 2.5 mm. In this case it is necessary to specify the thickness and total cross-section. The impact tests were carried out for at least 3 samples. The V–notch was cut in the weld from the face side. A load corresponding to impact energy of 150 J was used for the tests.

Samples for bending test were 20.0 mm wide and two samples were taken from each tested weld joint. The bending mandrel with a diameter of 35 mm was chosen according to the standard for a plate thickness of 3.0 mm and elongation of the steel A = 8%, Table 2.

The samples for the static tensile test were 25.0 mm wide in the gripping part, 12.0 mm wide and 70 mm long in the measuring part. The jaw travel speed was 0.008 mm/s.

3. Results and Discussion

3.1. Testing the Shape of Fusion Zone and the Quality of Bead-on-Plate Welds

The preliminary tests were conducted for the simulation of an autogenous laser welding process of butt joints of 3.0 mm thick Armox 500T steel plates. The aim of this stage of the study was to establish the range of basic laser welding parameters for butt joints, providing full penetration and accepted quality of the welds. The influence of the laser beam focal position on the penetration depth and shape of the fusion zone during bead-on-plate laser welding at laser output power was 1.25 kW, with a welding speed of 0.5 m/min, and an energy input of 150 J/mm (Figure 2). As can be seen, the geometry of the welds produced with the laser beam focused on the top surface (Figure 2a), and with the beam focused 1.5 mm under the top surface (Figure 2c) was almost identical, Table 3. In this case, the width of the weld face was 2.4–2.5 mm, while the width of the weld root was 1.3 mm. Therefore, the location of the focal plane in the range from the top surface to the center of the plate thickness (focal position from 0 to −1.5 mm) did not affect significantly the shape of the fusion zone and penetration depth. The shape of the fusion zone of both welds was in X configuration (also called the hourglass shape), and the depth/width ratio of FZ 1.25 indicates that the welds were produced at keyhole mode welding. Further lowering the focal position led to lack of penetration, and at the same time widening of the weld face to 2.7 mm as can be seen in Figure 2d. Focusing the laser beam above the top surface of the plate was also unfavorable, as it led to a change in shape to a Y configuration (also called as mushroom shape) and a reduction in width of the root, Figure 2b. In this case, the width of the weld face increased to approx. 3.0 mm, while the width of the weld root was reduced to only 0.3 mm. Moreover, the clear tendency of porosity formation mainly in the lower region of the welds was found in the range of the investigated parameters of welding, regardless of whether there was a full penetration or not.

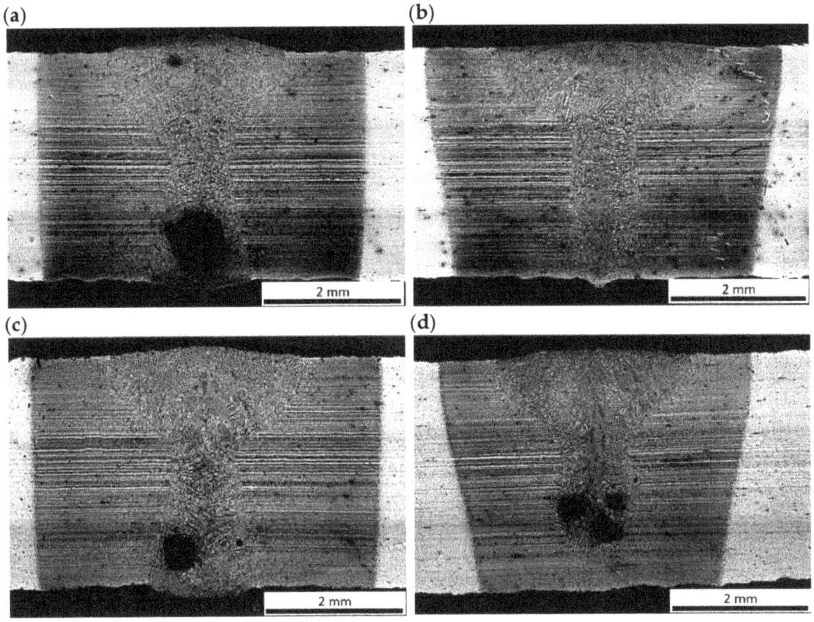

Figure 2. Influence of focal position of the laser beam on the penetration depth and shape of the fusion zone during bead-on-plate laser welding at laser output power 1.25 kW, welding speed 0.5 m/min (energy input 150 J/mm, Table 3); focal position: (**a**) B2, f: 0 mm, (**b**) B3, f: 1.5 mm, (**c**) B4, f: −1.5 mm, and (**d**) B5, f: −3.0 mm.

The influence of laser output power increase by 50% on the penetration depth and shape of the fusion zone during bead-on-plate laser welding at welding speed 0.5 m/min is shown in Figure 3. Laser welding at the laser output power 1.0 kW and welding speed 0.5 m/min, simply calculated via the energy input (ignoring the heat transfer coefficient) 120 J/mm, resulted in a weld face width of 2.0 mm but also lack of penetration, Figure 3a. Increasing the laser output power to 1.5 kW provided full penetration of the steel plate and resulted in an increase in the width of the weld face to 2.6 mm. The width of the root was 1.8 mm in this case, Figure 3b. While it is obvious that the increase in laser power resulted in an increase in heat input of welding and therefore, led to increase in penetration and width of the fusion zone, it should be noted that the energy input is a simplified parameter and does not directly determine the shape of the weld, especially in the case of laser welding in keyhole mode. This phenomenon is illustrated in Figure 4 which shows different geometry of welds produced at the same energy input 90 J/mm. The weld shown in Figure 4a, produced at a laser output power 1.5 kW and a welding speed 1.0 m/min, exhibits a Y shape fusion zone, with a weld face width of 2.7 mm and a root width of 1.2 mm. While the weld shown in Figure 4b, produced at double the laser output power (3.0 kW) and a welding speed (2.0 m/min), exhibits an X shape fusion zone, with a weld face width of 1.8 mm and a root width of 2.5 mm. Due to the significant difference in welding speed, despite constant energy input, the mechanisms of material heating, melting, solidification, and heat transfer are different. Moreover, the power density of the laser beam, as a heat source, affects the absorptivity of laser radiation and thus the efficiency of heat transfer. Therefore, the simply calculated energy input (based on power and welding speed) is not a determinant of the real heat input, which depends also on the efficiency of heat transfer.

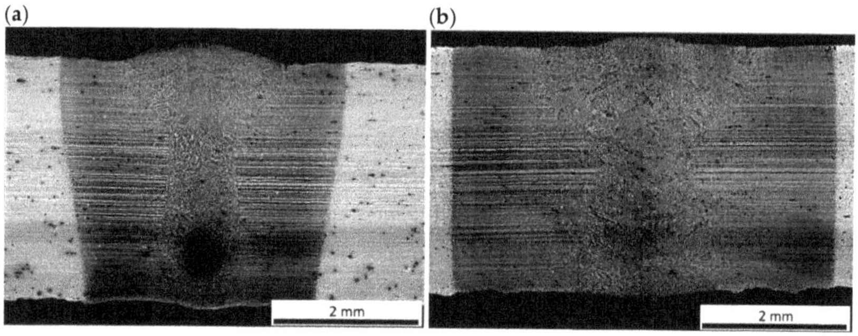

Figure 3. Influence of laser output power on the shape of fusion zone during bead-on-plate laser welding at welding speed 0.5 m/min and laser beam focused on the top surface (focal position, f 0 mm, Table 3); laser output power: (**a**) B1, 1.0 kW (energy input 120 J/min), (**b**) B6, 1.5 kW (energy input 180 J/min).

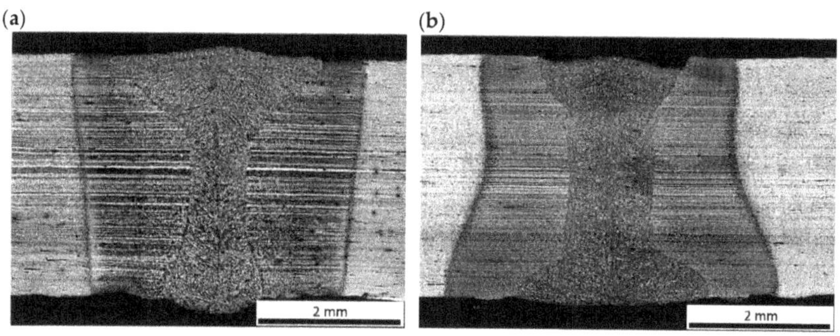

Figure 4. Comparison of the shape of fusion zone of welds produced during bead-on-plate laser welding at constant energy input of 90 J/mm, but different laser output power and welding speed (Table 3); (**a**) B7, laser output power 1.5 kW, welding speed 1.0 m/min; and (**b**) B8, laser output power 3.0 kW, welding speed 2.0 m/min.

The results of bead-on-plate laser welding also showed that the Y shape of welds and low welding speed of 0.5 m/min favored porosity mainly in the lower area of the weld (root porosity), Figures 2–4, and Table 3. This type of porosity is typical of laser welding in keyhole mode. Such porosity occurs usually within a certain range of welding parameters, and this phenomenon is related to the convection flow of liquid metal and partial evaporation. Many research results on the phenomenon of weld porosity during keyhole laser welding have been presented so far [18,21–23]. Some of the researchers found that the size and type of weld porosity (e.g., uniform, transitional, or root) is related mainly to the laser welding speed. In turn, other researchers pointed out that the porosity depends on the weld geometry (depth/width ratio). However, the geometry of the weld (fusion zone) during laser welding is related to the power density of the laser beam on the material (laser beam power relative to the surface area on which it affects, W/cm^2), and thus spot size, focal plane position, and also heat input of laser welding (both welding speed and laser power). Under such conditions the unstable keyhole may easily collapse. Therefore, the collapsing molten metal from the top regions of the keyhole is thought to trap the gases and vapors in the solidifying weld leading finally to pore formation in the weld metal [22–24,30–32].

3.2. Microstructure and Hardness Distribution of Butt Joints

The microstructure of the base metal of 3.0 mm thick Armox 500T steel plate is shown in Figure 5. As can be seen in Figure 5a, the base metal is characterized by very fine grains with clear segregation shearing lines, expanding along the plate rolling direction, as a result of inhomogeneous plastic deformation, in subsequent rolling passes. Moreover, the microphotograph in Figure 5a may suggest the presence of dispersed fine precipitations. In turn, based on the chemical composition, it can be suspected that they were carbides. However, the results of the XRD analysis of the base metal of Armox 500T steel at delivery conditions revealed peaks only from Feα' (martensite), see Figure 6. However, it should be noted that the detection level of the applied XRD method was about 3% for the specific γ phase. Therefore, it did not allow for the precise identification of the fine precipitations or phases at a share lower than 3%.

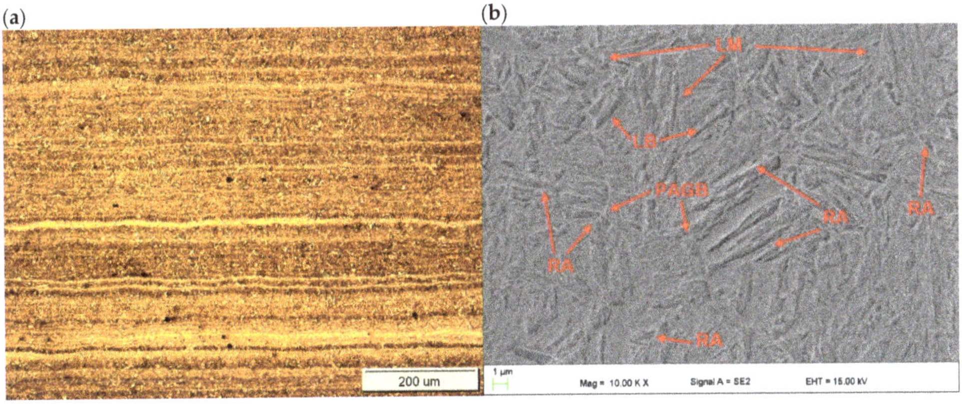

Figure 5. Microstructure of base metal of 3.0 mm thick Armox 500T steel plate (Tables 1 and 2); (**a**) optical micrograph, (**b**) SEM micrograph; LM—lath martensite, PAGB—prior austenite grain boundary, LB—lath boundaries, RA—traces of retained austenite.

Moreover, detailed observation of the microstructure at high magnification on SEM micrographs indicated that the base metal consisted mainly of tempered lath martensite. Some traces of morphology typical for retained austenite can be also observed, as shown in Figure 5b [1,3,33,34]. Planimetric analysis of the morphology typical of retained austenite indicated its share of 1.5–2%. The typical microstructure of Armox 500T steel consists of martensite and a few percent of retained austenite, as provided by the manufacturer (SSAB), and proved by many studies [1,3,4,11].

However, it should be noted that the microstructure of the armor steel Armox 500T is highly dependent on the heat treatment conditions. The microstructure changes with decreasing cooling rate during hardening, from martensitic to martensitic-bainitic, bainitic, bainitic-ferritic-pearlitic, to pearlitic-ferritic [3–5].

The fine-grained tempered lath martensite microstructure with a small amount of retained austenite is most preferable from the point of view of ballistic performance [1,4,6].

Hard tempered martensite with fine morphology and with some fraction of stable retained austenite is able to absorb excess of strain energy through plastically induced transformation into bcc-martensite. The most preferable amount of retained austenite in the case of armor steel plates should be in the range of 3–5%, and additionally it should be stable. Heat treatment, which leads to the precipitation of carbides and causes a significant reduction in ballistic resistance which was described by El-Fawakhry et al. [3].

Based on the initial tests of bead-on-plate laser welding, the extreme parameters with the highest and the lowest energy input, providing full penetration and required weld quality with limited tendency to porosity, were chosen for autogenous butt joint laser welding. The cross-sections of representative butt joints are presented in Figure 7. The butt joint produced at the lowest energy input of 70 J/mm shows just a slight undercut of the weld face (Figure 7a).

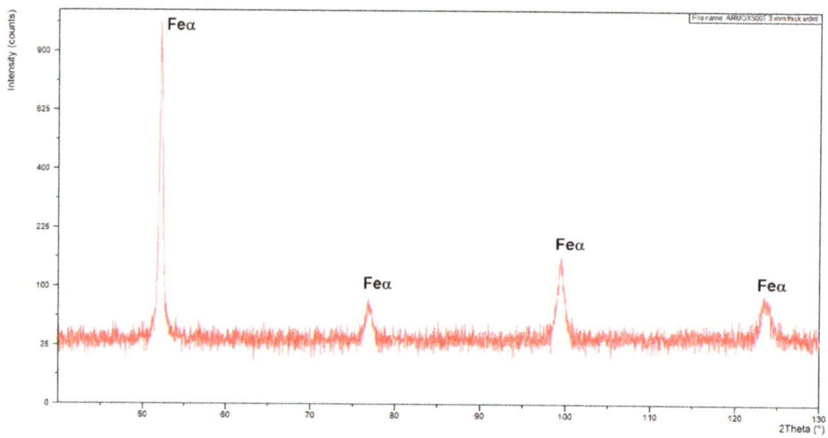

Figure 6. XRD pattern of the base metal Armox 500T steel plate 3.0 mm thick.

However, such weld face undercuts of not more than 10% of the plate thickness are typical and acceptable in the case of autogenous laser welding. In turn, the butt joint produced at the highest energy input 180 J/mm showed small single pores in the middle region of the weld, and with a diameter of not more than 125 µm (Figure 7b). The shape and the width of the weld and heat-affected zone are obviously related to the energy input. It is worth noting that the shape (configuration of fusion zone) and dimensions of the butt joints differed to the bead-on-plate welds. This phenomenon is related to the different conditions of melting and solidification of I-joint type during autogenous welding (without filler).

The optical micrographs of the central region of fusion zone, along with the fusion line and HAZ of the butt joints are shown in Figure 8. As can be seen, the fusion line was sharp and clear. Therefore, the coarse-grained region of HAZ, adjacent to the fusion line, was very narrow in the case of both butt joints. In this region, due to high temperature, a slight grain growth occurred if compared to the microstructure of the base metal. Due to recrystallization the austenite grains transformed into martensite with possible content of bainite and retained austenite. The XRD spectra of weld metal taken from both joints produced at different parameters were very similar with peaks only from Feα′ (martensite), as can be seen in Figure 9. However, traces of the morphology typical for retained austenite in weld metal can be also observed in SEM micrographs, as shown in Figures 10b and 11b.

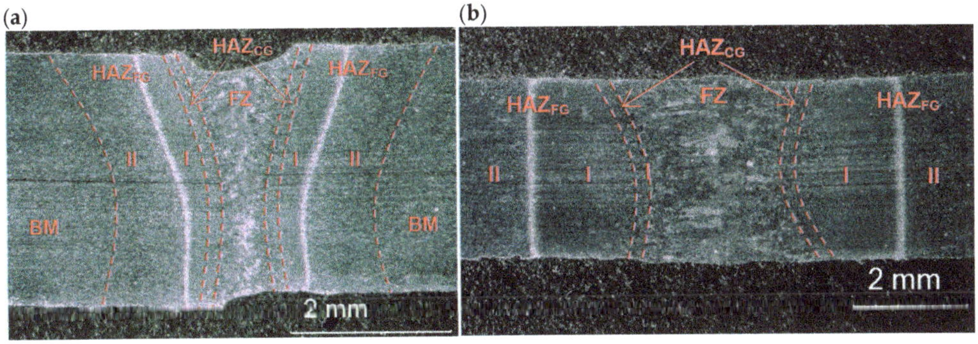

Figure 7. Cross-sections of the 3.0 mm thick butt joints of Armox 500T steel produced via autogenous laser welding (Table 4); (**a**) at lowest energy input of 70 J/mm (3.0 kW, 2.5 m/min), (**b**) at highest energy input of 180 J/mm (1.5 kW, 0.5 m/min); BM: Base Metal, FZ: Fusion Zone, HAZ_{FG}: Fine-Grained Heat-Affected Zone, HAZ_{FG}: Coarse-Grained Heat-Affected Zone.

Table 4. Parameters of autogenous butt joint laser welding of the 3.0 mm thick plates of Armox 500T steel.

Weld Bead No.	Output Laser Power kW	Welding Speed m/min	Focal Position * mm	Energy Input J/mm	Width of the Weld Face mm	Width of the Weld Root	Fusion Zone Shape	Remarks
BJ1	1.5	0.5	0	180	2.6	2.8	X	FP
BJ2	3.0	2.5	0	70	2.5	1.3	Y	FP,U

Remarks: * position relative to the top surface of plate; Quality assessment: LF—lack of fusion, FP—full penetration, U—undercut of wed face.

In turn, the fine-grained region and partially transformed region of HAZ was directly dependent on the energy input, as can be seen in Figures 7 and 8.

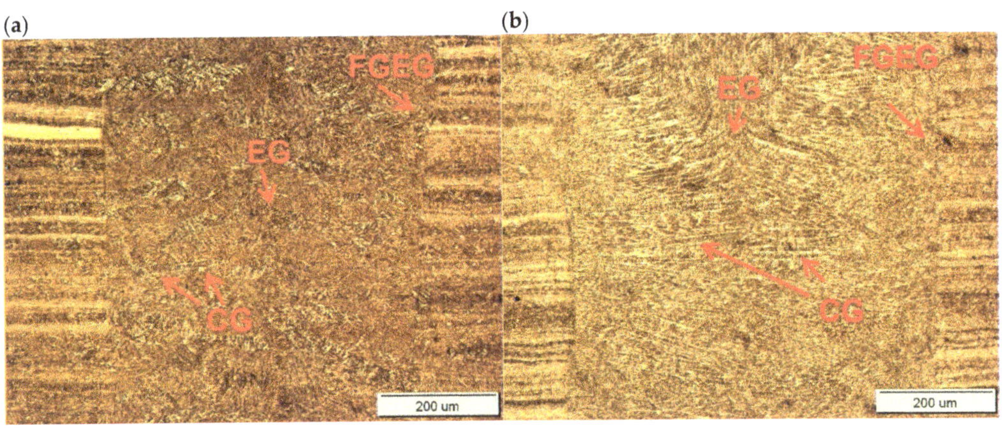

Figure 8. Optical micrographs of the middle region of fusion zone of 3.0 mm thick butt joints of Armox 500T steel produced via autogenous laser welding (Table 4); (**a**) at lowest energy input of 70 J/mm (3.0 kW, 2.5 m/min), (**b**) at highest energy input of 180 J/mm (1.5 kW, 0.5 m/min); CG: Columnar Grains, EG: Equiaxed Grains, FGEG: Fine-Grained Equiaxed Grains.

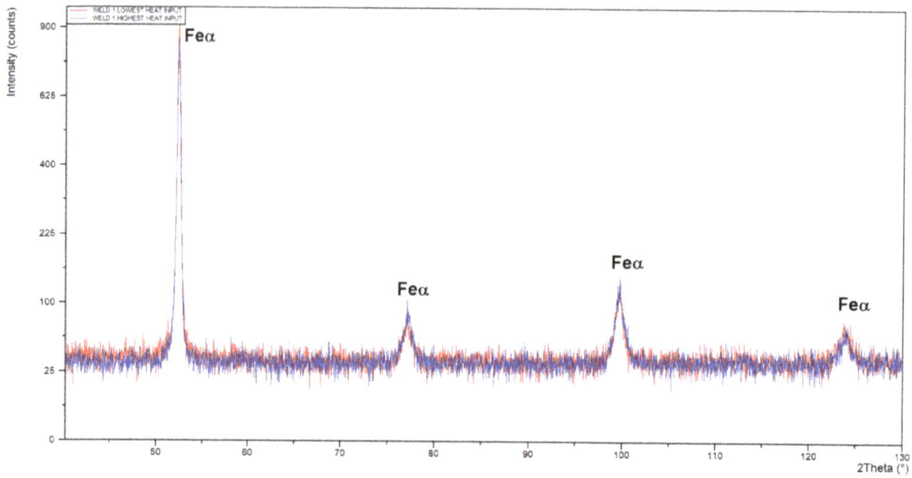

Figure 9. XRD pattern of butt joints of Armox 500T 3.0 mm thick steel produced via autogenous laser welding.

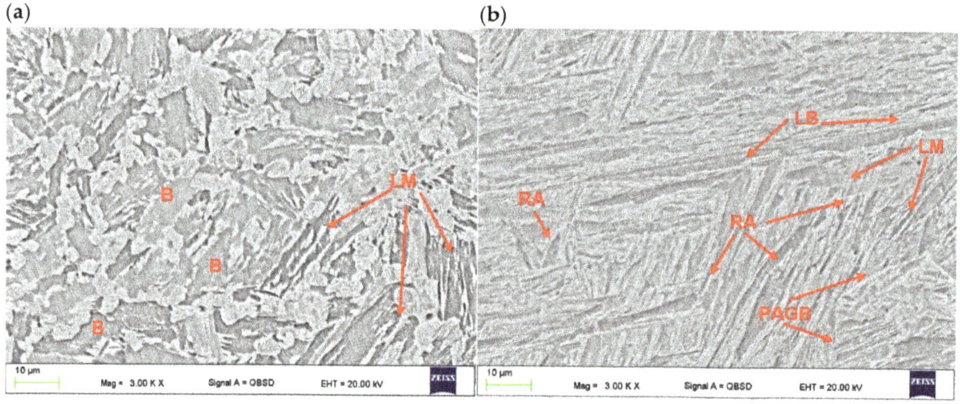

Figure 10. SEM micrographs of butt joints produced the at lowest energy input of 70 J/mm (Table 4); (**a**) HAZ adjacent to the fusion line, (**b**) fusion zone in the center region of weld; LM—lath martensite, PAGB—prior austenite grain boundary, LB—lath boundaries, B—bainite, RA—traces of retained austenite.

It should be noted that the HAZ did not end at the clearly visible "white line", but extended much further, as marked with red dashed lines in Figure 7a,b. The actual width of the HAZ was identified during measurements of microhardness distribution on the cross-section of the test joints. Additionally, the fine-grained region of HAZ was divided into two areas marked as I and II, due to the different distribution of microhardness. As can be see, the total width of the HAZ was twice to three times larger than the width of the fusion zone.

The SEM micrographs taken from the "white line" zone of fine-grained HAZ of butt joints produced at different parameters is shown in Figures 10a and 11a.

As can be seen, the microstructure of this zone differed from the base metal and fusion zone, and the share of martensite was much smaller and its morphology was different (small grained martensite), see Figures 10a and 11a. Other researchers also point to the existence of a soft zone with soft patches of white ferrite phase, placed between the fine-grained HAZ and base metal [6,18]. However, accurate microhardness measurements on the cross-section of the joints allowed for the precision location of areas with significant changes in hardness. The area of lowest hardness, where the hardness drops off rapidly, is the clearly visible "white line" in the fine-grained HAZ. In turn, between the fusion line and the "white line", the hardness even increases. SEM images of the microstructure in the "white line" zone located in the fine-grained HAZ of both tested test joints indicated the dominant share of bainite in relation to martensite, see Figures 10a and 11a.

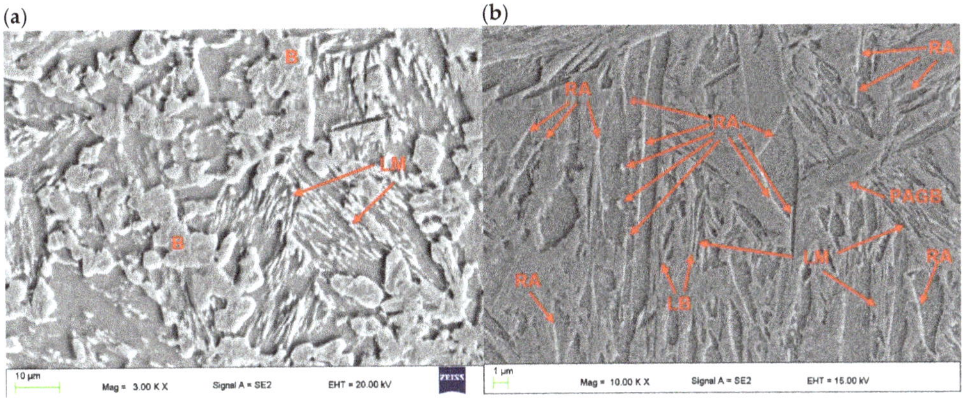

Figure 11. SEM micrographs of butt joints produced the at highest energy input of 180 J/mm (Table 4); (**a**) HAZ adjacent to the fusion line, (**b**) fusion zone in the center region of weld; LM—lath martensite, PAGB—prior austenite grain boundary, LB—lath boundaries, B—bainite, RA—traces of retained austenite.

The optical micrographs of the joints in the central region of welds exhibited typical primary structure with columnar grains growing from partially melted grains at the fusion line and equiaxed grains in the central region of the fusion zone (Figure 8). The structure of the fusion zone and the share of columnar and equiaxed grains obviously depended on the thermal conditions and heat input during laser welding.

The primary structure of the weld was formed during its solidification. The weld produced at the higher heat input exhibited a thin, fine-grained zone of equiaxed grains in the vicinity of the fusion line, which grew on partially melted grains of the base metal. The cooling rate during solidification in this region was the highest and the molten metal was highly supercooled. The width of this zone was approx. 40–50 µm, as can be seen in Figure 8b. When this zone was formed, it reduced the cooling rate, which reduced supercooling and the rate of nucleation as well. Therefore, this led to the formation of a zone of large columnar grains with a privileged crystallographic orientation, and the direction of their rapid growth coincided with the direction of heat dissipation. The solidification of the central zone of the weld took place with the lowest supercooling and without a clear direction of heat dissipation. Randomly oriented nuclei passed through the dendritic crystallization phase and grew to size depending on the thermal conditions of the welding process. As can be seen in Figure 8b, the columnar grain size is approx. 250–300 µm, which almost converged in the middle of the weld, due to which the area of equiaxed grains in its region is narrow.

On the other hand, the structure of the weld made with lower energy input was different, mainly due to the lower proportion of columnar grains, as can be seen in Figure 8a. This was related to the higher rate of cooling, crystallization, and thus the difference in the primary structure. It should be also noted that the structure on the cross-section of the weld was not homogeneous. The results of EDS analysis of some constituents of the weld metal of the joint produced at higher energy input (180 J/mm) are presented in Figure 12, Table 5. However, the higher cooling rate affected the extent and intensity of the martensitic transformation in the solid state, and thus the final microstructure of the weld. Moreover, the size of martensite laths in the grains also depended on the heat input. Higher heat input led to coarsening of the laths of martensite, as can be seen in Figures 10b and 11b.

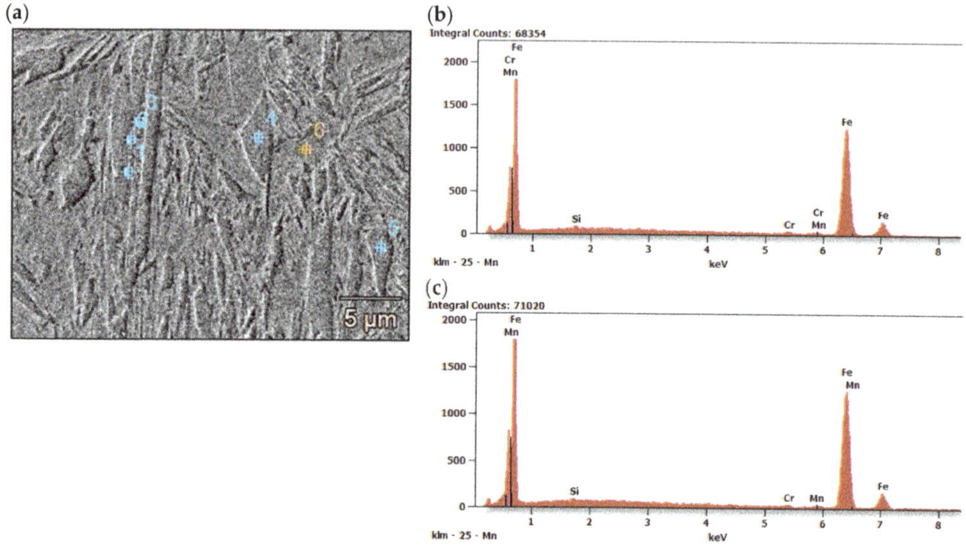

Figure 12. SEM micrographs of fusion zone of butt joint produced the at highest energy input of 180 J/mm (**a**), representative EDS spectra taken from point 1 (**b**) and point 4 (**c**).

Table 5. Quantitative results of EDS analysis in different points in fusion zone of butt joint produced the at highest energy input (Figure 12a).

Tested Region	Si		Cr		Mn		Fe	
	Wt.%	At.%	Wt.%	At.%	Wt.%	At.%	Wt.%	At.%
Point 1	0.4	0.8	0.9	1.0	1.4	1.4	97.3	96.8
Point 2	0.4	0.7	1.1	1.1	1.1	1.2	97.4	97.0
Point 3	0.3	0.7	1.0	1.0	1.1	1.1	97.6	97.2
Point 4	0.3	0.6	0.7	0.7	0.9	1.0	98.1	97.7
Point 5	0.4	0.9	0.8	0.9	1.8	1.9	96.9	96.4
Point 6	0.5	1.1	0.9	1.0	1.3	1.3	97.2	96.6

Microhardness measurements conducted on cross-sections of the butt joints showed that the base metal of Armox 500T steel was characterized by the microhardness ranging in 470–510 HV0.2 (Figure 13).

The microhardness distribution across the joint welded at lower energy input of 70 J/mm was very uniform with low scatter. The microhardness of the fusion zone was slightly lower compared to the BM from 477 to 480 HV0.2 (Figure 13). The highest values up to 526 HV0.2 were determined in the first region "I" of HAZ, see Figures 7a and 13a. As can be seen in Figure 13a, the rapid drop of the microhardness to375–390 HV0.2 occurred exactly at the sharp "white line". Then the microhardness increased gradually in the second region "II" of HAZ up to the microhardness of the base metal. A similar microhardness distribution occurred in the case of the joint welded at higher energy input of 180 J/mm. However, the width of the characteristic zones was larger, the microhardness values were different and the scatter of results was greater. It is worth noting that the microhardness was measured along three parallel lines, but the error bars were omitted so as not to reduce the clarity of the graphs, especially for the joint welded with higher energy input. As can be seen, the microhardness in the fusion zone varied in range from 446 HV0.2 to 525 HV0.2 (Figure 13b). There was also significant scatter of result in the coarse-grained HAZ adjacent to the fusion line. In this region the microhardness varied from 437 HV0.2 to 517 HV0.2 (Figure 13b). Similar to the previous joint, the maximum value of microhardness 531 HV0.2 was determined in the first region "I" of HAZ. The drop in microhardness at the "white line" zone was even greater and reached 315–345 HV0.2 (Figure 13b). Although, the decrease was significant, the microhardness level was typical of bainitic structure rather than ferritic structure.

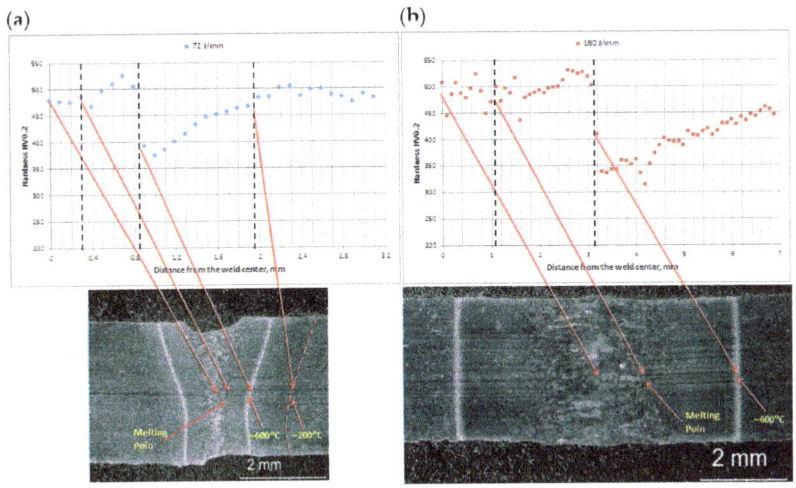

Figure 13. Microhardness distribution on the cross-section of 3.0 mm thick butt joints of Armox 500T steel produced via autogenous laser welding (Table 4); (**a**) at lowest energy input of 70 J/mm (3.0 kW, 2.5 m/min), (**b**) at highest energy input of 180 J/mm (1.5 kW, 0.5 m/min).

The hardness distribution was obviously related to the change in microstructure caused by the thermal cycle, i.e., the temperature range and cooling rate. Determining the temperature distribution in the HAZ under laser welding conditions is not easy due to the high temperature gradient and high traveling speed of the heat source. However, it can be assumed that at the fusion line the temperature reaches the melting point, as indicated in Figure 13. On the other hand, the influence of heat treatment of Armox 500T steel on hardness, tensile strength, and impact strength is well studied and there are many articles in this area. Hyeon-Seok et al. [35] found that there is no significant decrease in hardness at the tempering temperature of 150–200 °C. In turn, at the tempering temperature of 600 °C, the hardness decreased to approximately 320 HV [35].

The sudden drop in microhardness in the HAZ can be explained after estimating the temperature distribution in this area. If it is assumed that the hardness in the HAZ is proportional to the tempering temperature of Armox 500T steel, then characteristic temperature can be assigned to the boundary lines, as shown in Figure 13. An increase in the tempering temperature of Armox 500T steel up to 600 °C generally leads to coarsening of the martensite laths, carbon diffusion and a decrease in the dislocation density in martensite, as well as carbide precipitation. In turn, in the HAZ area, where the temperature is between 700 °C up to the fusion line, recrystallization of the primary structure takes place.

In the earlier work Lisiecki A. [15] proposed a methodology for determining the cooling time in the temperature range of 800 and 500 °C ($t_{8/5}$), i.e., in the HAZ recrystallization zone under autogenous laser welding conditions. Since the detailed methodology for determining the cooling time $t_{8/5}$ is presented in this reference [15], it was omitted in the current work. The cooling times determined according to this methodology for the butt joints of Armox 500T were 0.45 s for the heat input of 70 J/mm (BJ2) and 2.98 s for the heat input of 180 J/mm (BJ1), respectively.

3.3. Mechanical Properties and Fracture Mode of Butt Joints

The results of technological bending test of butt joints of 3.0 mm thick Armox 500T steel autogenously laser welded revealed low plasticity of the welds and HAZ of the joints. While the base metal of Armox 500T steel should withstand a full bending angle of 180°, the butt joint's failure occurred at a bending angle of just 25–35°, Figure 14, Table 6.

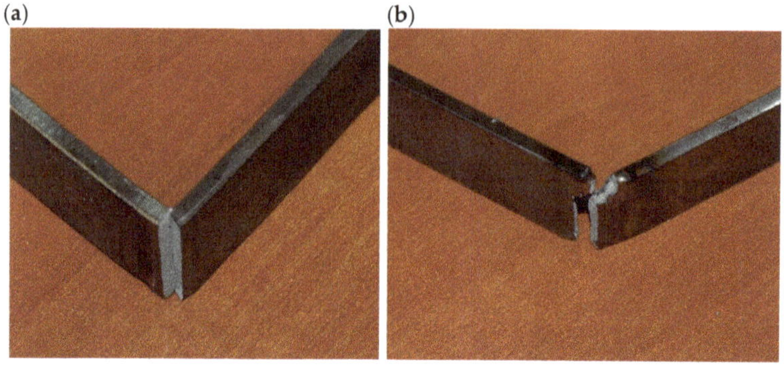

Figure 14. A view of the samples after technological bend test (weld root bending) of 3.0 mm thick butt joints of Armox 500T steel produced via autogenous laser welding (Tables 4 and 6); (**a**) at lowest energy input of 70 J/mm (3.0 kW, 2.5 m/min), (**b**) at highest energy input of 180 J/mm (1.5 kW, 0.5 m/min).

Table 6. Results of the technological bending test of 3.0 mm thick butt joint of Armox 500T steel produced via autogenous laser welding (Table 4), according to the PN-EN ISO 5173:2023-06 standard, Figure 14.

Joint/Sample Type	Test Type	Bending Angle, °	Remarks/Location of Failure
JA (70 J/mm)	Face bend	35	Longitudinal and transverse fracture along the entire width/HAZ/FZ
	Root bend	32	Longitudinal fracture along the entire width/HAZ
JB (180 J/mm)	Face bend	25	Longitudinal fracture along the entire width/HAZ
	Root bend	28	Longitudinal fracture along the entire width/HAZ

Remarks: HAZ—heat-affected zone, FZ—fusion zone.

The results of static tensile test of butt joints of 3.0 mm thick Armox 500T steel autogenously laser welded has shown satisfactory ultimate tensile strength, especially in the case of the joint welded at the highest energy input, although the rupture occurred at the HAZ location with the lowest hardness. The mean value of tensile strength for samples of the joint welded at 180 J/mm was over 1400 MPa (Figure 15 and Table 7).

On the other hand, the joint welded at the lowest energy input of 70 J/mm showed lower tensile strength, which was related to the crack location at the fusion zone boundary (Figure 15 and Table 7). However, in this case, the zone between the fusion line and the region of the lowest microhardness was very narrow, approximately 0.5 mm, so it was not possible to clearly identify the place where the crack was initiated. Additionally, a decrease in hardness was also identified near the fusion line, which may indicate a decrease in strength in this area, Figure 13a.

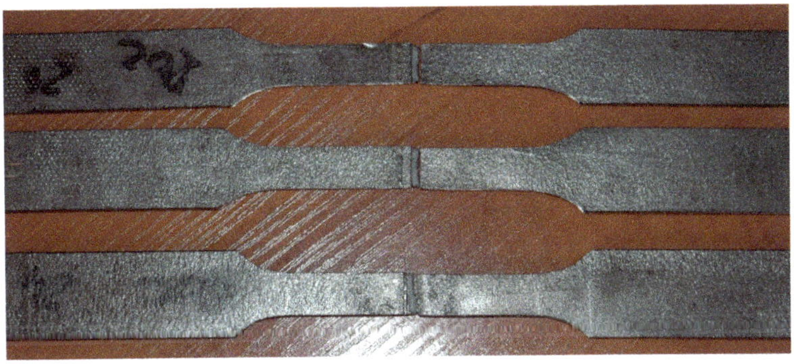

Figure 15. A view of broken samples after the static tensile tests of butt joint of Armox 500T steel produced via autogenous laser welding (Table 4) at lowest energy input of 70 J/mm.

Table 7. Results of static tensile test of 3.0 mm thick butt joint of Armox 500T steel produced via autogenous laser welding (Table 4), according to the PN-EN ISO 148-1:2017-02 standard, Figure 15.

Joint Type	Sample No.	Tensile Strength Rm, MPa	Location of Failure
JA (70 J/mm)	JA1	1373.09	FZ/HAZ
	JA2	1369.14	FZ/HAZ
JB (180 J/mm)	JB1	1402.39	HAZ (weakest zone)
	JB2	1401.29	HAZ (weakest zone)

Remarks: HAZ—heat-affected zone, FZ—fusion zone, BM—base metal.

The hardness drop and weakening of the material in HAZ was related with coarsening the microstructure due to thermal cycle, especially in the region of coarse-grained HAZ but also as a result of tempering and the related structure transformation resulted in the formation of bainitic microstructure in a certain zone; this can be seen as a white line in the HAZ. Moreover, in the case of the investigated steel the thermal cycle in HAZ may lead to the transformation of the retained austenite into bainite, which is less ductile, and also to precipitation of complex carbides such $M_{23}C_6$ and M_7C_3 [24,30–32,34,36,37].

The results of Charpy V–notch test of the butt joints 3.0 mm thick of Armox 500T steel produced via autogenous laser welding at different energy inputs, tested at room temperature, and also at reduced temperature are summarized in Table 7. As can be seen, the impact toughness of the welded joints tested at 20 °C (room temperature) was at the level required for the base metal of Armox 500T steel at −40 °C, Tables 2 and 7. In turn, the impact toughness of the welded joints tested at −20 °C is approximately 25–30% lower than values determined at room temperature.

The fracture surface of the Charpy impact samples were examined under SEM microscope to determine the mode of fracture of butt joints produced at different heat input and different temperature. As can be seen in Figures 16b and 17b, the fracture surface of samples tested at −20 °C showed mainly cleavage fracture mode for both joints. The cleavage facets and very limited amount of dimples, as well as microvoids and holes can be observed on the fracture surface in the case of joint produced at the lowest energy input of 70 J/mm (Figure 16b). In turn, the fracture surface of the joint produced at the higher energy input of 180 J/mm shows rather quasi-cleavage facets with clear shearing areas and tear ridges (Figure 17b). This also corresponded with the higher impact energy absorbed during the tests by this joint (9.5 J), compared to the joint produced at lower energy input (8.6 J), as seen in Table 8.

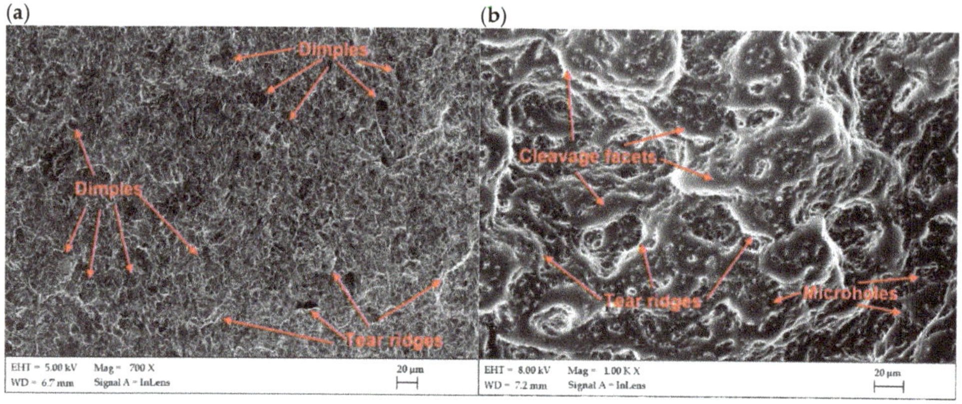

Figure 16. SEM images of fracture surfaces of Charpy impact samples of butt joints produced the at lowest energy input of 70 J/mm (Table 4); (**a**) tested at 20 °C, (**b**) tested at −20 °C.

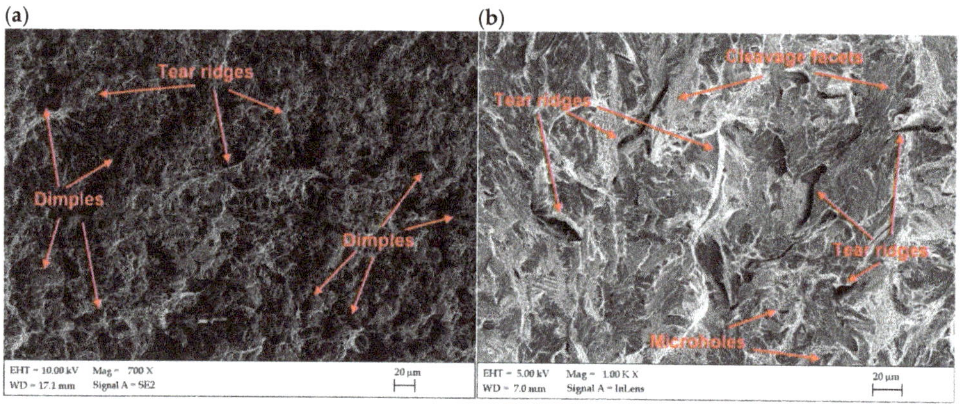

Figure 17. SEM images of fracture surfaces of Charpy impact samples of butt joints produced the at lowest energy input of 180 J/mm (Table 4); (**a**) tested at 20 °C, (**b**) tested at −20 °C.

Table 8. Results of Charpy V–notch test of 3.0 mm thick butt joint of Armox 500T steel produced via autogenous laser welding (Table 4), according to the PN-EN ISO 4136:2022-12 standard, see Figures 16 and 17.

Joint Type	Temperature of Charpy V–Notch Test	Mean Value of Absorbed Impact Energy, J	Mean Value of Impact Toughness, J/cm^2
JA (70 J/mm)	20 °C	11.4	34.2
	−20 °C	8.6	25.8
JB (180 J/mm)	20 °C	13.3	39.9
	−20 °C	9.5	28.5

In turn, the samples tested at 20 °C (room temperature) indicated a mixed mode of fracture with dimples that can be seen on the surfaces. Such fracture surface topography indicated that the progress of damage partially followed a void growth and subsequent coalescence (Figure 16a,b). The fracture surfaces in Figures 16a and 17a showed a large density of the fine dimples of different sizes. Some microvoids and holes can be also identified on fracture surfaces of both tested joints. However, more ductile dimples may be observed on the fracture surface of the joint produced at higher energy input. The results of impact test also showed higher impact energy absorbed by the joint (13.3 J), compared to the joint produced at a lower energy input (11.4 J).

In general, the drop of impact toughness of weld metal was mainly related to the grain size coursing and grain orientation [24,30,33]. Considering that higher energy input of laser welding leads to grain growth, it can be concluded that the grains orientation had a greater influence on the impact characteristics of the tested butt joints produced within the specific parameters and conditions of welding.

4. Conclusions

Based on the results of the study on autogenous laser welding of 3.0 mm thick Armox 500T plates, the following conclusions were drawn:

- Low welding speed of 0.5 m/min and Y-type shape of fusion zone favor porosity in the bottom of the weld (root porosity), which is a common problem during keyhole laser welding.
- Welding at higher speed provided a columnar (I configuration) or hourglass (X configuration) shape of the fusion zone, and eliminated the risk of porosity. Therefore, it was possible to ensure satisfactory quality of butt joints, characterized by a proper geometry, and no internal defects.
- Microstructure of the fusion zone and HAZ was related directly to the energy input of laser welding, thus thermal conditions of solidification and cooling rates. The dominant phase constituent in the fusion zone and HAZ was martensite. The fusion zone consisted of columnar and equiaxed grains with the size of martensite dependent on the energy input of laser welding process.
- The butt joints were characterized by significantly lower microhardness in HAZ due to transformation of microstructure into bainitic in the specific zone. Therefore, the static tensile strength of the joints was lower that the base metal. Impact tests conducted at reduced temperature (−20 °C) showed a reduction of the impact toughness by approx. 25–30%, if compared to the values achieved at room temperature. The fracture surface of samples tested at −20 °C exhibited clearly a cleavage fracture mode, while the samples tested at 20 °C exhibited a mixed mode of fracture with some dimples.
- Narrow weld and HAZ may be beneficial from the point of view of the ballistic performance of the butt joint of armor plates.
- It is planned to continue research in the field of laser welding greater thicknesses of armor plates.

Author Contributions: Conceptualization, methodology, investigation, formal analysis, validation, supervision, writing—review and editing, A.L.; fractography and microstructure analysis, A.K.-L.; SEM, EDS analysis, W.P.; formal analysis and support, G.C., G.F.B. and M.A. All authors have read and agreed to the published version of the manuscript.

Funding: This research received no external funding.

Institutional Review Board Statement: Not applicable.

Data Availability Statement: The original contributions presented in the study are included in the article, further inquiries can be directed to the corresponding author.

Acknowledgments: Authors thank W. Kwaśny for his help with the phase analysis.

Conflicts of Interest: Author Grzegorz Chrobak was employed by the company IPG Photonics Sp. z o.o. The remaining authors declare that the research was conducted in the absence of any commercial or financial relationships that could be construed as a potential conflict of interest.

References

1. Fei, Z.; Pan, Z.; Cuiuri, D.; Li, H.; Gazder, A.A. A Combination of Keyhole GTAW with a Trapezoidal Interlayer: A New Insight into Armour Steel Welding. *Materials* **2019**, *12*, 3571. [CrossRef] [PubMed]
2. Jena, P.; Mishra, B.; Rameshbabu, M.; Babu, A.; Singh, A.; Sivakumar, K.; Bhat, T.B. Effect of heat treatment on mechanical and ballistic properties of a high strength armour steel. *Int. J. Impact Eng.* **2010**, *37*, 242–249. [CrossRef]
3. El-Fawakhry, M.K.; Fathy, A.M. The Effect of Bad Heat Treatment Technology on Failure Mode of Armor Steel Sheet under EN1522 Ballistic Test. *Int. J. Mater. Technol. Innov. (IJMTI)* **2021**, *1*, 30–44. [CrossRef]
4. Starczewski, L.; Szczech, S.; Tudyka, D. Tests of armour steels in view of their protective efficiency. *Prace IMŻ* **2010**, *62*, 110–117.
5. Khan, W.; Tufail, M.; Chandio, A.D. Characterization of Microstructure, Phase Composition, and Mechanical Behavior of Ballistic Steels. *Materials* **2022**, *15*, 2204. [CrossRef] [PubMed]
6. Magudeeswaran, G.; Balasubramanian, V.; Madhusudhan Reddy, G. Effect of welding consumables on hydrogen induced cracking of armour grade quenched and tempered steel welds. *Ironmak. Steelmak.* **2008**, *35*, 549–560. [CrossRef]
7. Turichin, G.; Kuznetsov, M.; Klimova-Korsmik, O.; Sklyar, M.; Zhitenev, A.; Kurakin, A.; Pozdnyakov, A. Laser-Arc hybrid welding perspective ultra-high strength steels: Influence of the chemical composition of weld metal on microstructure and mechanical properties. *Procedia CIRP* **2018**, *74*, 752–756. [CrossRef]
8. Kumar, S.N.; Balasubramanian, V.; Malarvizhi, S.; Rahman, A.H.; Balaguru, V. Effect of welding consumables on shielded metal arc welded ultra high hard armour steel joints. *J. Mech. Behav. Mater.* **2022**, *31*, 8–21. [CrossRef]
9. Skowrońska, B.; Szulc, J.; Bober, M.; Baranowski, M.; Chmielewski, T. Selected properties of RAMOR 500 steel welded joints by hybrid PTA-MAG. *J. Adv. Join. Process.* **2022**, *5*, 100111. [CrossRef]
10. Górka, J.; Stano, S. Microstructure and Properties of Hybrid Laser Arc Welded Joints (Laser Beam-MAG) in Thermo-Mechanical Control Processed S700MC Steel. *Metals* **2018**, *8*, 132. [CrossRef]
11. Janicki, D. Disk laser welding of armor steel. *Arch. Metall. Mater.* **2014**, *59*, 1641–1646. [CrossRef]
12. Bassett, J. Laser welding of high hardness armour steel. *Sci. Technol. Weld. Join.* **1998**, *3*, 244–248. [CrossRef]
13. Grajcar, A.; Morawiec, M.; Rozanski, M.; Stano, S. Twin-spot laser welding of advanced high-strength multiphase microstructure steel. *Opt. Laser Technol.* **2017**, *92*, 52–61. [CrossRef]
14. Sowards, J.W.; Hussey, D.S.; Jacobson, D.L.; Ream, S.; Williams, P. Correlation of Neutron-Based Strain Imaging and Mechanical Behavior of Armor Steel Welds Produced with the Hybrid Laser Arc Welding Process. *J. Res. Nat. Inst. Stand. Technol.* **2018**, *123*, 1. [CrossRef] [PubMed]
15. Lisiecki, A. Welding of Thermomechanically Rolled Fine-Grain Steel by Different Types of Lasers. *Arch. Metall. Mater.* **2014**, *59*, 1625–1631. [CrossRef]
16. Kurc-Lisiecka, A.; Lisiecki, A. Laser welding of new grade of advanced high strength steel DOMEX 960. *Mater. Tehnol.* **2017**, *51*, 199–204. [CrossRef]
17. Kurc-Lisiecka, A. Impact toughness of laser-welded butt joints of the new steel grade Strenx 1100MC. *Mater. Tehnol.* **2017**, *51*, 643–649. [CrossRef]
18. Landowski, M.; Świerczyńska, A.; Rogalski, G.; Fydrych, D. Autogenous Fiber Laser Welding of 316L Austenitic and 2304 Lean Duplex Stainless Steels. *Materials* **2020**, *13*, 2930. [CrossRef] [PubMed]
19. Kurc-Lisiecka, A.; Lisiecki, A. Weld Metal Toughness of Autogenous Laser-Welded Joints of High-Strength Steel Domex 960. *Mater. Perform. Charact.* **2019**, *8*, 20190071. [CrossRef]
20. Kurc-Lisiecka, A.; Lisiecki, A. Hybrid Laser-GMA Welding of High-Strength Steel Grades. *Mater. Perform. Charact.* **2019**, *8*, 20190070. [CrossRef]
21. Lisiecki, A. Effect of Heat Input During Disk Laser Bead-On-Plate Welding of Thermomechanically Rolled Steel on Penetration Characteristics and Porosity Formation in The Weld Metal. *Arch. Metall. Mater.* **2016**, *61*, 93–102. [CrossRef]
22. Norris, J.T.; Robino, C.V.; Hirschfeld, D.A.; Perricone, M.J. Effects of Laser Parameters on Porosity Formation: Investigating Millimeter Scale Continuous Wave Nd:YAG Laser Welds. *Weld. J.* **2011**, *90*, 198–203.
23. Torkamany, M.J.; Sabbaghzadeh, J.; Hamedi, M.J. Effect of laser welding mode on the microstructure and mechanical performance of dissimilar laser spot welds between low carbon and austenitic stainless steels. *Mater. Des.* **2012**, *34*, 666–672. [CrossRef]
24. Pandey, C.; Saini, N.; Mahapatra, M.; Kumar, P. Study of the fracture surface morphology of impact and tensile tested cast and forged (C&F) Grade 91 steel at room temperature for different heat treatment regimes. *Eng. Fail. Anal.* **2017**, *71*, 131–147.
25. *PN-EN ISO 17637:2017-02*; Non-Destructive Testing of Welded Joints—Visual Testing of Welded Joints. Polish Committee for Standardization: Warszawa, Poland, 2017.
26. *PN-EN ISO 9015-2:2016-04*; Destructive Testing of Welded Metal Joints—Hardness Testing—Part 2: Microhardness Testing of arc Welded Joints. Polish Committee for Standardization: Warszawa, Poland, 2016.
27. *PN-EN ISO 5173:2023-06*; Destructive Testing of Welds in Metal Materials—Bending Testing. Polish Committee for Standardization: Warszawa, Poland, 2023.
28. *PN-EN ISO 148-1:2017-02*; Metals—Charpy Impact Test—Part 1: Test Method. Polish Committee for Standardization: Warszawa, Poland, 2017.
29. *PN-EN ISO 4136:2022-12*; Destructive Testing of Welded Metal Joints—Tensile Test of Transverse Samples. Polish Committee for Standardization: Warszawa, Poland, 2022.

30. Dak, G.; Sirohi, S.; Pandey, C. Study on microstructure and mechanical behavior relationship for laser-welded dissimilar joint of P92 martensitic and 304L austenitic steel. *Int. J. Press. Vessel. Pip.* **2022**, *196*, 104629. [CrossRef]
31. Wang, Z.Q.; Wang, X.L.; Nan, Y.R.; Shang, C.J.; Wang, X.M.; Liu, K.; Chen, B. Effect of Ni content on the microstructure and mechanical properties of weld metal with both-side submerged arc welding technique. *Mater. Charact.* **2018**, *138*, 67–77. [CrossRef]
32. Almeida, D.T.; Clarke TG, R.; de Souza JH, C.; Haupt, W.; de Lima MS, F.; Mohrbacher, H. The effect of laser welding on microstructure and mechanical properties in heavy-gage press hardening steel alloys. *Mater. Sci. Eng. A* **2021**, *821*, 141341. [CrossRef]
33. Guo, W.; Cai, Y.; Wang, B.; Mu, W.; Xin, D. Effect of microstructure synergism on cryogenic toughness for CGHAZ of low-carbon martensitic steel containing nickel. *Mater. Sci. Eng. A* **2022**, *830*, 142240. [CrossRef]
34. Kozłowska, A.; Janik, A.; Radwański, K.; Grajcar, A. Microstructure Evolution and Mechanical Stability of Retained Austenite in Medium-Mn Steel Deformed at Different Temperatures. *Materials* **2019**, *12*, 3042. [CrossRef]
35. Hyeon-Seok, L.; Jimin, L.; Young-Beum, S.; Hong-Kyu, K.; Byoungchul, H. Effect of Tempering Temperature on the Microstructure and Mechanical Properties of ARMOX 500T Armor Plate. *Korean J. Mater. Res.* **2017**, *27*, 359–363. [CrossRef]
36. Adamiak, M.; Tomiczek, B.; Górka, J.; Czupryński, A. Joining of the AMC Composites Reinforced with Ti_3Al Intermetallic Particles by Resistance Butt Welding. *Arch. Metall. Mater.* **2016**, *61*, 847–852. [CrossRef]
37. Adamiak, M.; Czupryński, A.; Kopyść, A.; Monica, Z.; Olender, M.; Gwiazda, A. The Properties of Arc-Sprayed Aluminum Coatings on Armor-Grade Steel. *Metals* **2018**, *8*, 142. [CrossRef]

materials

MDPI

Article

Multi-Response Optimization of Aluminum Laser Spot Welding with Sinusoidal and Cosinusoidal Power Profiles Based on Taguchi–Grey Relational Analysis

Saeid SaediArdahaei and Xuan-Tan Pham *

Department of Mechanical Engineering, École de Technologie Supérieure, 1100, Notre-Dame Ouest Street, Montreal, QC H3C 1K3, Canada; saeid.saediardahaei.1@ens.etsmtl.ca
* Correspondence: tan.pham@etsmtl.ca; Tel.: +1-(514)-396-8468

Abstract

Laser weld quality remains a critical priority across nearly all industries. However, identifying optimal laser parameter sets continues to be highly challenging, often relying on costly, time-consuming trial-and-error experiments. This difficulty is largely attributed to the severe fluctuations and instabilities inherent in laser welding, particularly keyhole instabilities. This study examines the impact of laser power modulation parameters, which, when properly applied, have been found effective in controlling and minimizing process instabilities. The investigated parameters include different pulse shapes (sinusoidal and cosinusoidal) and their associated characteristics, namely frequency (100–800 Hz) and amplitude (1000–4000 W). The impact of these modulation parameters on keyhole mode laser spot welding performance in aluminum is investigated. Using a Taguchi experimental design, a series of tests were developed, focusing on eight key welding responses, including keyhole dimensions, mean temperature, and the variability of instability-inducing forces and related factors affecting process stability. Grey relational analysis (GRA) combined with analysis of variance (ANOVA) is applied to identify the optimal combinations of laser parameters. The results indicate that low amplitude (1000 W), low to intermediate frequencies (100–400 Hz), and cosinusoidal waveforms significantly enhance weld quality by improving process stability and balancing penetration depth. Among the factors, amplitude has the greatest impact, accounting for over 50% of the performance variation, followed by frequency and pulse shape. The findings provide clear guidance for optimizing laser welding parameters to achieve stable, high-quality aluminum welds.

Keywords: keyhole; surface tension; darcy damping force; instability; optimization; RSD; GRA analysis; ANOVA

1. Introduction

Aluminum and its alloys are widely used in industry due to their high strength-to-weight ratio, corrosion resistance, and excellent electrical conductivity [1]. Being nearly three times lighter than steel and about half the weight of titanium, they are particularly suitable for electric vehicle body structures and battery components [2].

Laser welding is widely adopted in the automotive and aerospace sectors for producing deep, precise welds in aluminum alloys [3,4]. However, challenges persist due to aluminum's high reflectivity, thermal conductivity, and oxidation tendency [5]. Pulsed wave (PW) lasers help address these issues by using high initial energy density to overcome reflectivity while maintaining controlled average power to reduce overall heat input [6,7].

Keyhole-mode laser welding, a deep penetration technique for aluminum alloys, has been widely studied due to its complex process control. This mode typically initiates at a power density around 10^6 W/cm^2, where intense metal vaporization creates a keyhole and induces recoil pressure [8]. Keyhole-mode laser welding enhances productivity with welding speeds up to 20–30 times faster, while optimized parameters improve the mechanical properties of welded aluminum components [9].

Porosity is a common challenge in aluminum welding, mainly due to hydrogen solubility and keyhole instability. The evaporation of volatile alloying elements (e.g., Zn, Mg, and Li) further destabilizes the process [5]. Additionally, the narrow process window and rapid keyhole formation within microseconds complicate experimental studies due to erratic weld pool and keyhole behavior [10]. Inherent instabilities in keyhole laser welding limit its broader industrial adoption. Keyhole stability depends on the dynamic interplay of highly transient and fluctuating forces, such as surface tension (Marangoni and curvature effects), recoil pressure, hydrostatic and hydrodynamic pressures, and Darcy damping. Balancing these forces is essential to improve stability and reduce defects [11,12].

Laser power manipulation has shown effectiveness in controlling instabilities. Ardahaei et al. provided insights into the numerical modeling of keyhole instabilities in laser spot welding of aluminum using power wave modulation (PWM) [12,13]. Their study revealed that curvature effects and Darcy damping forces are key contributors to keyhole instability. Optimized pulse shapes, such as single- or quadruple-peak triangular and ramp-down rectangular pulses, effectively reduced spontaneous force fluctuations, enhancing process stability and minimizing defect probability. In a separate study, they demonstrated that rectangular pulses increased penetration depth by over 80% compared to continuous wave (CW) lasers, with power and frequency modulation significantly affecting keyhole behavior. Matasunawa et al. [14] showed that PWM considerably minimizes porosity by stabilizing keyhole and molten pool behavior. Tsukamoto et al. [15] further demonstrated that aligning modulation frequency with the molten pool's natural oscillations significantly enhances keyhole stability and minimizes porosity formation. Zhang et al. [16] investigated sinusoidal power modulation in fiber laser welding of AZ31B magnesium alloy, reporting improved energy coupling, reduced underfill, refined microstructure, and enhanced mechanical properties due to improved keyhole and melt pool stability. Heider et al. [17] studied copper welding and found that sinusoidal power modulation, particularly at normalized frequencies of 0.2–0.4, reduced defects like pores and melt ejections by up to 90%, while improving keyhole stability. They optimized parameters such as modulation frequency, amplitude, and focal size to achieve deeper, more stable penetration.

A review of the literature indicates a gap in multi-response optimization for keyhole-mode laser welding of aluminum. While several researchers have proposed such techniques, most studies have lacked a comprehensive approach to simultaneously optimize multiple desired outcomes. Omoniyi et al. [18] applied the Taguchi method combined with grey relational analysis (GRA) to optimize the laser welding of Ti6Al4V, balancing bead geometry and micro-hardness. Their study highlighted GRA's effectiveness in multi-objective welding problems where single-response optimization is insufficient. In another research, Tsai et al. [19] used Taguchi-GRA to optimize multiple quality attributes, such as roundness, taper, and HAZ, in the laser drilling of acrylic. By integrating diverse metrics into a single grey relational grade, the method effectively identified optimal parameters and balanced trade-offs among competing objectives, demonstrating GRA's strength in multi-response optimization. This study presents a numerical investigation of keyhole-mode laser spot welding on aluminum, experimentally validated and optimized using Taguchi-based GRA. A novel cosinusoidal PWM profile was introduced and compared with continuous and sinusoidal profiles to assess its effectiveness in minimizing keyhole instabilities. GRA was employed to address the lack of robust multi-response optimization methods, enhancing simulation precision by reducing factors contributing to instability.

2. Experimental Details and Materials

Figure 1 shows the experimental setup for laser spot welding on aluminum using continuous, sinusoidal, and cosinusoidal pulse profiles. A fiber pulsed laser machine (IPG YLS-6000: Ytterbium Laser, IPG Photonics Corporation, Marlborough, MA, USA) with a maximum output laser power of 6 kW and a laser wavelength of 1090 nm was utilized to perform laser spot welds on test samples. The focal length of the laser was 300 mm, producing a beam diameter of 0.3 mm. The laser system used in this study was manufactured by IPG Photonics Corporation, headquartered in Marlborough, Massachusetts, United States. The material being used was Aluminum 6061 with a 2 mm thickness. A series of continuous, sinusoidal, and cosinusoidal laser pulses were used to create laser spots on the material. The surface was brushed prior to welding to remove contaminants and prevent oxidation. Due to aluminum's high reflectivity, a 7.5 degree deviation was applied to the laser head to avoid damaging the welding head. The thermal properties of aluminum 6061 are listed in Table 1.

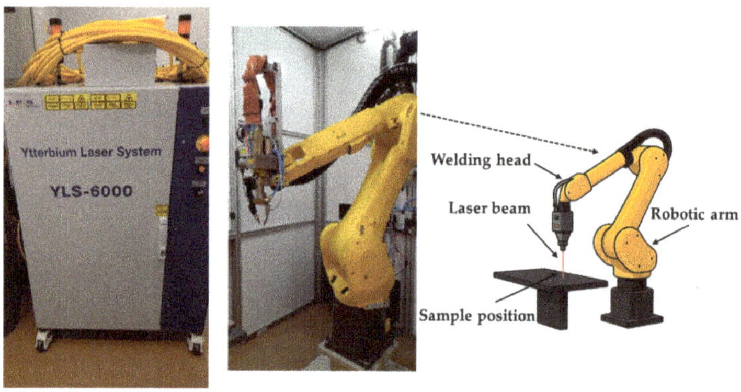

(**a**) (**b**)

Figure 1. Experimental setup of the laser welding system: (**a**) laser source (IPG YLS-6000: Ytterbium Laser), (**b**) laser welding head and components.

Table 1. Thermophysical properties of 6061-T6 aluminum alloy [20,21].

Property	Symbol	Magnitude
Solidus temperature	T_s	873.13 (K)
Liquidus temperature	T_l	915.15 (K)
Vaporization temperature	T_V	2760 (K)
Thermal conductivity of solid	k_s	235 (W/m/K)
Thermal conductivity of liquid	k_l	90 (W/m/K)
Density of solid	ρ_s	2660 (kg/m^3)
Density of liquid	ρ_l	2380 (kg/m^3)
Latent heat of melting	L_m	3.87×10^5 (J/kg)
Latent heat of vaporization	L_V	1.05×10^7 (J/kg)
Specific heat capacity of solid	$C_{p,s}$	870 (J/kg/K)
Specific heat capacity of liquid	$C_{p,l}$	1170 (J/kg/K)

Table 1. *Cont.*

Property	Symbol	Magnitude
Convective heat transfer coefficient	h	20 (W/m^2/K)
Coefficient of linear thermal expansion	β	2.8×10^{-5} (1/K)
Dynamic viscosity	μ	1.3×10^{-3} (Pa.s)
Coefficient of surface tension	σ	$0.95 \times (1 + 0.13 \times (1 - T/T_m))^{1.67}$ (N/m)
Temperature-dependent surface tension coefficient	$\partial\sigma/\partial T$	-0.15×10^{-3} (N/m/K)
Radiation emissivity	ξ	0.2

3. Methods and Configurations

Laser spot welding was simulated using a 2D axisymmetric model in COMSOL Multiphysics 5.6, justified by the rotational symmetry of the stationary laser beam around the vertical *z*-axis. This setup allowed for efficient representation of 3D behavior via symmetry. This study investigated the pulse modulation impact by applying sinusoidal and cosinusoidal laser power distributions, focusing on keyhole stability and overall welding efficiency. The study applied sinusoidal and cosinusoidal laser power profiles to examine their impact on keyhole stability and welding efficiency. The computational domain and power profiles are shown in Figure 2. Boundaries ABGF and EGCD represent air and aluminum, respectively. A Gaussian beam was used to model the laser heat source.

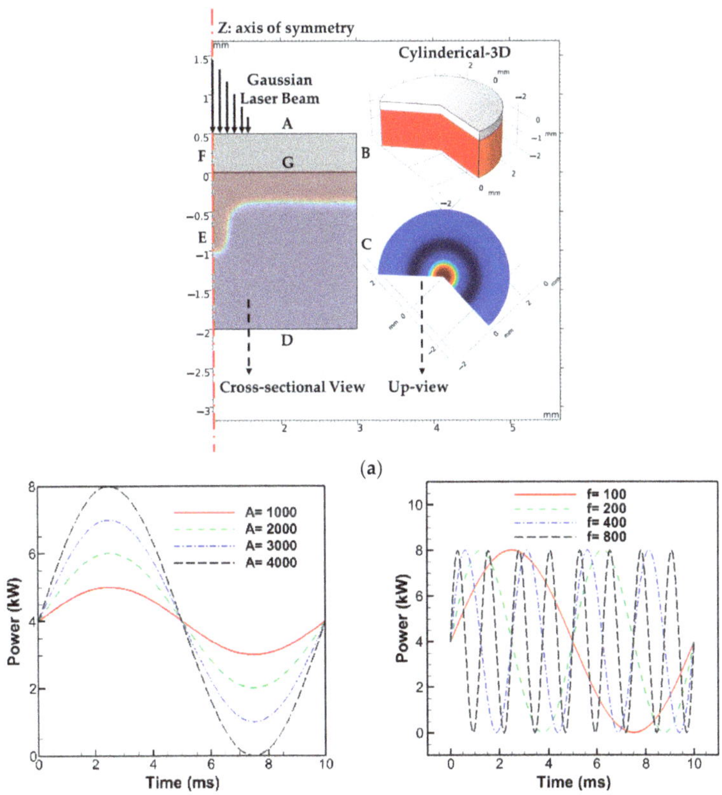

Figure 2. *Cont.*

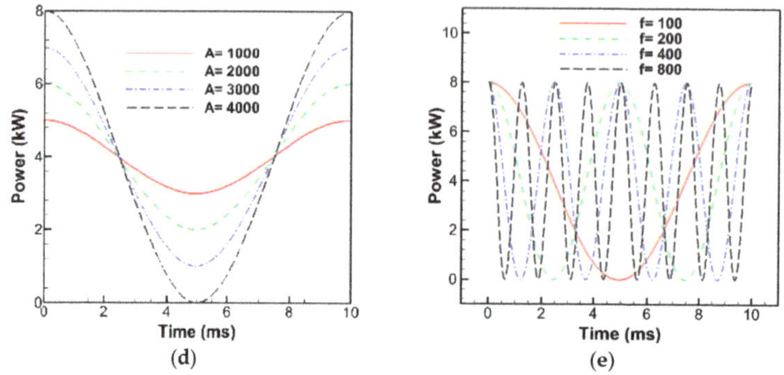

Figure 2. (**a**) Schematic 2D axisymmetric representation of numerical sample geometry and results in COMSOL Multiphysics. (**b**,**c**) Examples of sinusoidal power profiles under constant and variable amplitudes/frequencies. (**d**,**e**) Examples of cosinusoidal power profiles under constant and variable amplitudes/frequencies.

3.1. Heat and Fluid Dynamics

A detailed description of the numerical methods, governing equations, and mesh sensitivity analysis used in this study is available and verified in our previous publications [12,13]. A brief summary is provided here, as the focus is on developing and applying a novel optimization technique. The modified level set (LS) method [22], modified mixture theory (MMT) [13,23], and the thermal enthalpy porosity technique (TEPT) [24,25] were used to model multiphase interactions during laser welding. These models were employed due to their proven effectiveness in dealing with phase transformations, keyhole morphology, and fluid flow. These approaches effectively capture vapor/liquid interactions (via LS/MMT) and solid/liquid interactions (via MMT/TEPT), ensuring reliable simulation results. MMT also facilitates finite element calculations across multi-phase elements applying mixture effects [13,23]. The model incorporated recoil pressure, surface tension forces, Darcy damping force, evaporation and mass loss, buoyancy, and surface dynamics. The numerical framework of this study was based on several assumptions to simplify the complex multiphase simulation: (a) Newtonian, incompressible laminar flow; (b) temperature-independent thermophysical properties; (c) mushy zone as a porous medium [12]; (d) neglecting multiple beam reflections; and (e) treating vaporized material as an ideal gas. The transport phenomena across all phases were modeled by solving the modified forms of energy (Equation (1)), mass (Equation (2)), and momentum (Equation (3)) conservation equations, and the LS (Equation (4)) equation.

$$\rho C_p \frac{\partial T}{\partial t} + \rho C_p \vec{u} \cdot \nabla T = \nabla \cdot (k \nabla T) + (q_{Laser} - Q_{vapor}) \delta(\phi) \quad (1)$$

$$\nabla \cdot \vec{u} = \delta(\phi) \dot{m}_{H-L} \left(\frac{\rho_l - \rho_v}{\rho^2} \right) \quad (2)$$

$$\rho \left(\frac{\partial \vec{u}}{\partial t} + \vec{u} \cdot \left(\vec{\nabla} \cdot \vec{u} \right) \right) = \vec{\nabla} \cdot [-pI + \mu(\vec{\nabla} \vec{u} + \left(\vec{\nabla} \vec{u} \right)^T)]) + \rho \vec{g} - \rho_l \beta_l (T - T_{melting}) \vec{g} \phi - \mu_l K \vec{V} + (\gamma \cdot nk - \nabla_s \gamma \cdot t) \delta(\phi) \quad (3)$$

$$\frac{\partial \phi}{\partial t} + \vec{u} \cdot \nabla \phi - \delta(\phi) \dot{m}_{H-L} \left(\frac{V_{f,1}}{\rho_v} + \frac{V_{f,2}}{\rho_l} \right) + \gamma_{ls} \nabla \cdot \left(\phi(1 - \phi) \frac{\nabla \phi}{|\nabla \phi|} - \epsilon_{ls} \nabla \phi \right) = 0 \quad (4)$$

The energy equation (Equation (1)) includes the velocity vector $\vec{u}$, specific heat capacity C_p, temperature T, thermal conductivity k, and time t. The laser energy heat source q_{laser}

is defined as $\left(2\alpha P_{Laser}/\,\pi R_{eff}^{2}\right)exp\left(-2r^{2}/R_{eff}^{2}\right)B_{t}$ where P_{Laser} is the peak laser power, B_t is the analytic pulse-shaping function, α is the aluminum absorptivity, and R_{eff} is the effective beam radius. Energy loss due to evaporation, Q_{vapor}, is calculated as $-L_V\dot{m}_{H-L}$, where L_V is the latent heat of vaporization, and $\dot{m}_{H-L}$ refers to the mass loss due to evaporation, defined as $(1-\beta_r)\sqrt{M/2\pi R}\left(P_{sat}(T)/\sqrt{T}\right)$. The saturated vapor pressure P_{sat} is determined by $P_{atm}exp[(1-(T_v/T))ML_v/RT_v]$ [26]. Here, β_r is the retro-diffusion coefficient, R the universal gas constant, and M the molar mass of vaporized particles. A delta function $\delta(\phi)$, characterized using the LS variable ϕ, applies the laser heat flux and energy loss due to evaporation only at the vapor/liquid interface (where $\phi = 0.5$). Equation (2) represents the modified mass conservation equation incorporating a source term to add the impact of recoil pressure [25,27]. Equation (3) represents the momentum conservation, including the dynamic viscosity μ, pressure and viscous stresses, gravity, buoyancy, Darcy damping, and surface tension forces. The Darcy drag coefficient K, is defined with a relation as $\left((180\,/\,d^2)\,(1-V_l)^2\right)/(V_l^{\,3}+b))$, with d, the dendrite size constant set to 10^{-2} cm [28], and b, the constant used to avoid division by zero. β_l and $\left(\vec{\nabla}\vec{u}\right)^T$ are the volume thermal expansion coefficient and the transpose of velocity vector gradient. V_l is the liquid volume fraction, assigned to one above the liquidus temperature, zero below the solidus temperature, and $(T-T_s)/(T_l-T_s)$ between the two [24,29,30]. Equation (4) is the modified LS equation, improved by adding a gas dynamic source term to model vaporization effects due to mass loss and vapor pressure at the vapor/liquid interface [22]. Parameters γ_{ls} and ϵ_{ls} are the reinitialization and the interface thickness values, respectively, set based on sensitivity analysis [13]. $V_{f,1}$ and $V_{f,2}$ denote the gas and solid/liquid volume fractions.

3.2. Optimization Approach

The Taguchi-based GRA was performed to optimize the numerical data from cosinusoidal and sinusoidal power profiles for enhancing keyhole laser welding performance on aluminum.

3.2.1. Taguchi Design

The Taguchi method enhances product quality through system, parameter, and tolerance design strategies. This study adopted parameter design, focusing on identifying robust parameter settings through controlled experimentation. An orthogonal array (OA) was used to structure the experimental layout, minimizing the number of simulations. The selection of OA was guided by computational cost, time constraints, and study objectives [31]. The total degrees of freedom (DOF) of the experiment are defined in Equation (5).

$$DOF_{experiment} = \sum DOF_{factor} + \sum DOF_{interactions} \tag{5}$$

Table 2 lists the factors and levels used in the Taguchi design. The interaction DOF between two factors was calculated as the product of their individual DOFs, where each factor's individual DOF was one unit less than its number of levels.

Table 2. Investigated factors and levels.

Symbol	Parameter	Level			
		1	2	3	4
A	Pulse Shape	Sinusoidal	Cosinusoidal	N/A	N/A
B	Frequency (Hz)	100	200	400	800
C	Amplitude (W)	1000	2000	3000	4000

This study considered three factor interactions: $A \times B$, $A \times C$, and $B \times C$. The total DOF yielded $[(2-1)+2\times(4-1)]_{DOF_{factor}} + [2\times(1\times 3)+(3\times 3)]_{\sum DOF_{interactions}} = 22$. A suitable OA must have a DOF equal to or greater than this value. Accordingly, an L_{32} OA was selected to fully capture all main and interaction effects influencing the keyhole welding process, as shown in Table 3.

Table 3. Chosen L32 mixed OA values.

Experiments	A	B	C
1	1	1	1
2	1	1	2
3	1	1	3
4	1	1	4
5	1	2	1
6	1	2	2
7	1	2	3
8	1	2	4
9	1	3	1
10	1	3	2
11	1	3	3
12	1	3	4
13	1	4	1
14	1	4	2
15	1	4	3
16	1	4	4
17	2	1	1
18	2	1	2
19	2	1	3
20	2	1	4
21	2	2	1
22	2	2	2
23	2	2	3
24	2	2	4
25	2	3	1
26	2	3	2
27	2	3	3
28	2	3	4
29	2	4	1
30	2	4	2
31	2	4	3
32	2	4	4

3.2.2. Grey Relational Analysis (GRA)

The primary objective of this study was to optimize the welding process by minimizing keyhole instability. Our previous work confirmed that large fluctuations in surface tension, Darcy damping force, and velocity significantly increase the likelihood of keyhole collapse by promoting instability [12]. Additionally, parameters such as mean temperature, keyhole depth, and width are critical for ensuring mechanical integrity and welding efficiency. Therefore, a multi-response optimization technique was developed to simultaneously account for these key parameters and their desired behaviors.

- Grey relational analysis (GRA)

The GRA approach is a technique that takes into account all the responses and combines them into one particular response by transforming them into Grey relational grades (GRGs) [18,19]. This study aimed to optimize desirable responses, such as keyhole depth, width, and mean temperature, while minimizing relative standard deviations (RSDs) of the Darcy damping force, surface tension, and velocity as key contributors to instability. The response objectives are summarized in Table 4. Preprocessing involved converting the raw data into a comparable format. The equations for calculating the standard deviation (SD) and RSD of the fluctuating forces and velocity are shown in Equations (6) and (7). The GRG computation followed three steps: normalization, Grey relational coefficient (GRC) calculation, and final GRG evaluation, with normalization performed using either Equation (8) or Equation (9), depending on the objective type being chosen as 'larger-the-better' or 'smaller-the-better' [32].

$$SD_i^{(i)} = \sqrt{1/n\sum_{t=1}^{n}\left(x_j^{(i)}(t) - \overline{x}_j^{(i)}\right)^2} \tag{6}$$

$$RSD_j^{(i)} = \left(SD_j^{(i)}/\overline{x}_j^{(i)}\right) \times 100 \tag{7}$$

$$X_{ij}{}^{*} = (X_{ij} - min(X_j))/(maxX_j - minX_j) \tag{8}$$

$$X_{ij}{}^{*} = (maxX_j - X_{ij})/(maxX_j - minX_j) \tag{9}$$

In Equations (6) and (7), $x_j^{(i)}(t)$ denotes the time-dependent value of the j-th response (e.g., surface tension, Darcy damping force) for the i-th test at time step t, and $\overline{x}_j^{(i)}$ represents its mean over all n = 1000 simulation steps sampled from t = 0 to t = 0.01 s. In Equations (8) and (9), X_{ij} and $X_{ij}{}^{*}$ are the original and normalized values of the j-th response of the i-th test. X_j is defined as a set of all values of the j-th response across all 32 experiments or test cases ($X_j = \{X_{1j}, X_{2j}, \ldots X_{32j}\}$). All the response values were scaled into [0, 1] using these equations (Equations (8) and (9)).

Table 4. Chosen L32 mixed orthogonal array values.

Parameter	Objective	Reason
Keyhole depth/width	Higher the better	Deeper welds
Mean temperature	Smaller the better	Less overall material heat exposure
RSD of Darcy damping force RSD of surface tension force RSD of velocity	Smaller the better	Reduced variability of parameter over time Fewer fluctuations; less instability

The next steps are devoted to calculating the deviation from the ideal normalized value and the GRC, which are shown in Equations (10) and (11) [32]. Finally, the GRG can be calculated using Equation (12).

$$\Delta_{ij} = \left|X_{ij}{}^{*,ideal} - X_{ij}{}^{*}\right| \tag{10}$$

$$GRC_{ij} = (\Delta_{min} + \xi\Delta_{max})/(\Delta_{ij} + \xi\Delta_{max}) \tag{11}$$

$$GRG_i = (1/m)\sum_{j=1}^{m} GRC_{ij} \tag{12}$$

where Δ_{ij} denotes the deviation of the normalized response from the ideal normalized response (typically taken as 1), while Δ_{min} and Δ_{max} represent the minimum and maximum possible deviations, usually set to 0 and 1. The distinguishing coefficient ξ, ranging from 0 to 1, was set to 0.5 for stability [18,32]. In Equation (12), m is the total number of responses considered in this analysis (here, m = 8).

- Analysis of variance (ANOVA)

The performance of the tests, factors, and levels in the Taguchi method was analyzed using a statistical measurement approach known as ANOVA. This quantified the contribution of each factor, their main effects, interactions, and associated errors on the overall response. Since the optimization data were obtained from single-run numerical simulations, standard ANOVA was applied without considering noise factors. The analysis assessed the statistical significance of each factor on the optimized response, represented by the grey relational grade (GRG).

3.3. System, Software, and Calculation Details

COMSOL Multiphysics 5.6 was used for simulations on a Lenovo ThinkStation P720 equipped with an Intel® Xeon® Gold 5118 CPU (12 cores, 24 threads) and 128 GB RAM. The CPU was produced by Intel corporation, based in Santa Clara, California, USA. The system was assembled by LENOVO, headquartered in Beijing, China. Three physics interfaces, including Heat Transfer in Fluids, Laminar Flow, and Level Set, were employed and coupled using the Non-Isothermal Flow and Two-Phase Flow interfaces. These ensured accurate modeling of three-phase phenomena by linking the heat transfer, fluid flow, and level set dynamics. A time step of 10 µs was used, with an extra-fine mapped mesh (0.02 mm quadrilateral elements) optimized for fluid dynamics. PARDISO solvers were applied: a nested dissection multithreaded version for fluid flow and an automatic preordering version for the Heat Transfer and Level Set equations. The optimal interface thickness and reinitialization parameter for the level set method were set to 5 m/s and 0.03 mm, respectively, ensuring better computational efficiency. All mathematical calculations related to the optimization technique were performed using Microsoft Excel.

4. Results and Discussion

This section presents the experimental and numerical results, optimization outcomes, and ANOVA analysis for laser spot welding on aluminum using sinusoidal and cosinusoidal power profiles. Eight key criteria were used to guide the optimization, with emphasis on minimizing keyhole instabilities and enhancing penetration depth and width. Parameters such as the mean temperature, RSDs of surface tension and Darcy damping forces, maximum fluid velocity, and keyhole geometry were analyzed to determine the optimal welding conditions.

4.1. Experimental Results and Validation

The effects of the sinusoidal and cosinusoidal laser power profiles were evaluated against the conventional continuous profile through laser spot welding trials. Cross-sectional and top-view images (Figures 3 and 4) illustrate that both the weld depth and width increased with modulated profiles. Specifically, the depth and width increased from 0.755 mm and 1.788 mm (continuous) to 0.900 mm and 1.985 mm (cosinusoidal, 2000 W amplitude), and further to 1.119 mm and 2.011 mm (sinusoidal, 2000 W amplitude). Moreover, increasing the amplitude from 1000 W to 2000 W at 100 Hz further enhanced the weld geometry. For the cosinusoidal profiles, the depth and width increased from 0.781 mm and 1.874 mm to 0.900 mm and 1.985 mm; for the sinusoidal profiles, they increased from 0.813 mm and 1.890 mm to 1.119 mm and 2.011 mm.

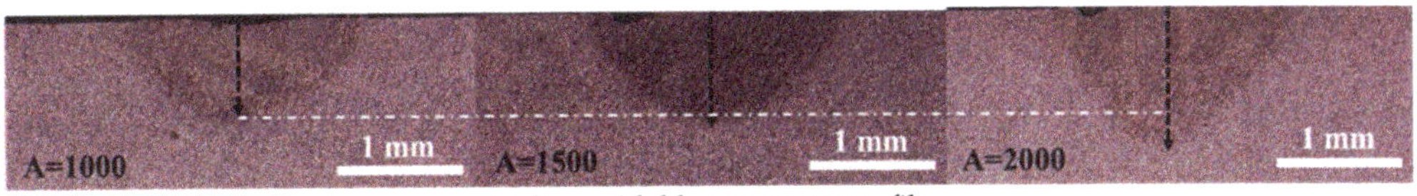

(**a**) Sinusoidal laser power profile.

(**b**) Continuous laser power profile.

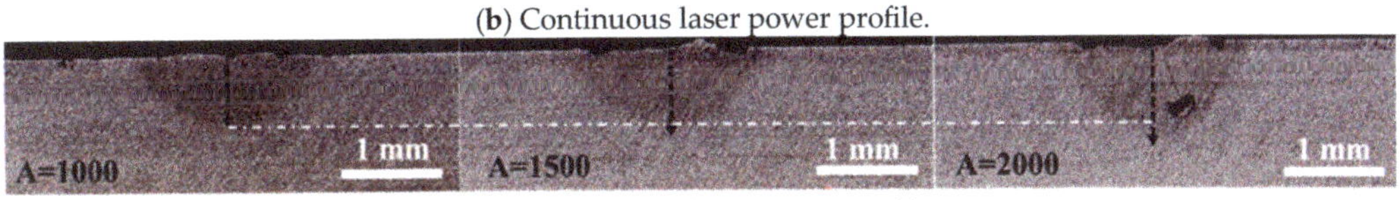

(**c**) Cosinusoidal laser power profile.

Figure 3. Cross-sectional view of the samples under different laser power profiles using an identical total laser energy of 40 j (4 kW over 10 ms) for (**a**) sinusoidal laser power profile under various laser power amplitudes, (**b**) continuous laser power profile, and (**c**) cosinusoidal laser power profile under various laser power amplitudes.

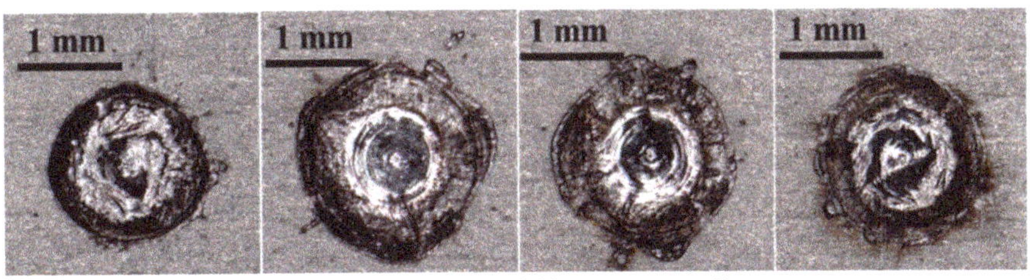

(**a**) CW versus sinusoidal laser power profile.

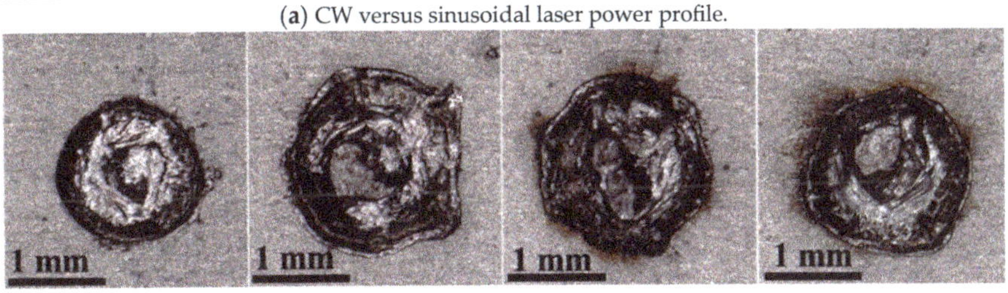

(**b**) CW versus cosinusoidal laser power profile.

Figure 4. Top view of the laser spots produced under CW and sin/cosinusoidal laser power profiles for (**a**) comparison of CW with sinusoidal laser power profiles with varying laser power amplitudes (left to right: CW, Sin: A = 2000, 1500, 1000) and (**b**) comparison of CW with cosinusoidal laser power profiles with varying laser power amplitudes (left to right: CW, Cos: A = 2000, 1500, 1000).

Figure 4 presents top-view images of the laser spots under different power profiles. The weld width increased with both sinusoidal and cosinusoidal profiles and further widened with higher power amplitudes compared to the continuous mode. These results confirm the superiority of modulated profiles in achieving greater depth-to-width ratios using the same total laser energy.

The numerical simulation was validated by comparing the simulated weld width and depth with the experimental results for three cases: the sinusoidal, cosinusoidal (100 Hz, 2000 W), and continuous power profiles. As shown in Figure 5, the simulated keyhole and molten pool dimensions closely matched the experimental measurements, with minimal error. This strong agreement confirms the reliability of the simulation approach in accurately predicting weld geometry and keyhole behavior.

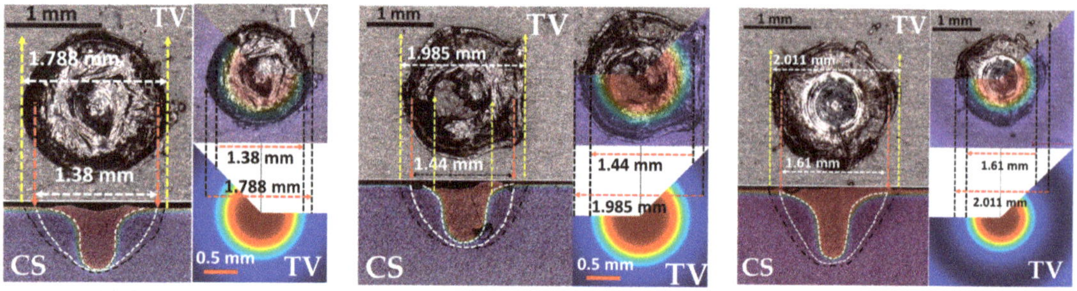

(**a**) Continuous. (**b**) Cosinusoidal. (**c**) Sinusoidal.

Figure 5. The numerical and experimental results of the top view and cross-section comparisons for (**a**) CW, (**b**) cosinusoidal (A = 2000, f = 100), and (**c**) sinusoidal (A = 2000, f = 100); the black dashed lines in the cross-section images represent the weld zone and HAZ, while the white dashed lines represent the numerically-derived weld zone and keyhole lines. CS is short for cross-section view, and TV is the top view.

4.2. Numerical Results and Optimization Procedures

Following the confirmed improvements in weld geometry using the sinusoidal and cosinusoidal profiles, the numerical study was extended to explore various frequencies and amplitudes under a constant total laser energy of 40 J for consistency. Using the Taguchi method (Tables 2 and 3), 32 simulations were performed. GRA and ANOVA were applied to determine the optimal welding parameters. The numerical results for the eight optimization criteria are summarized in Table 5 and Figure 6, with GRA-related and ANOVA data presented in Tables 6 and 7.

Table 5. The obtained test results for eight response variables under varying frequencies and amplitudes (RSD $Sur_{r,z}$: RSD of surface tension in the radial or axial direction).

Test	Keyhole Depth (mm)	Keyhole Width (mm)	RSD Velocity (%)	Mean Temperature (K)	RSD $Darcy_r$ (%)	RSD $Darcy_z$ (%)	RSD Sur_r (%)	RSD Sur_z (%)
1	0.793	1.500	46.9	2682	62.4	56.8	42.4	23.1
2	0.930	1.610	53.2	2403	58.0	72.0	54.8	36.9
3	1.067	1.670	56.8	2112	74.4	62.6	62.0	39.8
4	1.218	1.728	52.8	1941	77.8	62.2	67.1	39.5
5	0.852	1.434	44.9	2684	60.3	52.5	35.3	22.3
6	0.951	1.516	48.8	2439	58.7	63.4	48.9	34.9
7	1.008	1.580	53.3	2152	71.5	58.0	54.4	38.9
8	1.116	1.616	47.9	1995	78.4	58.2	57.8	38.9
9	0.848	1.384	46.4	2692	58.5	42.8	27.0	23.0
10	0.880	1.446	54.4	2543	63.2	55.6	43.8	28.6
11	0.945	1.488	62.4	2298	62.9	57.3	50.9	36.7
12	0.997	1.508	59.8	2135	75.2	55.4	53.0	37.6
13	0.784	1.350	53.9	2698	56.9	39.8	23.6	27.4
14	0.777	1.396	58.2	2665	59.1	46.9	35.6	26.8
15	0.801	1.444	54.5	2531	67.4	59.2	46.5	28.7
16	0.886	1.502	62.0	2358	67.2	59.9	50.0	31.1
17	0.754	1.350	30.6	2695	53.3	45.7	28.2	23.7
18	0.776	1.436	29.5	2382	54.1	49.8	51.2	35.7
19	0.857	1.474	29.5	2114	73.4	43.9	56.5	34.5
20	0.984	1.514	32.2	1948	75.1	41.7	59.7	33.5

Table 5. *Cont.*

Test	Keyhole Depth (mm)	Keyhole Width (mm)	RSD Velocity (%)	Mean Temperature (K)	RSD $Darcy_r$ (%)	RSD $Darcy_z$ (%)	RSD Sur_r (%)	RSD Sur_z (%)
21	0.762	1.404	42.8	2697	55.0	40.5	29.9	25.0
22	0.784	1.492	44.6	2444	56.4	56.3	49.0	34.1
23	0.881	1.536	52.0	2162	73.5	49.8	56.2	36.5
24	1.062	1.576	54.1	1999	73.4	48.3	60.9	35.0
25	0.810	1.396	45.2	2697	53.4	39.6	26.7	21.8
26	0.903	1.442	53.0	2548	59.6	55.0	43.8	27.1
27	0.995	1.496	62.2	2293	69.4	57.4	52.8	35.7
28	1.076	1.528	64.6	2125	74.2	58.5	57.7	37.4
29	0.839	1.360	54.6	2689	51.2	41.2	25.0	25.8
30	0.863	1.380	58.7	2641	54.4	48.4	39.9	24.8
31	0.882	1.420	57.4	2505	63.1	61.8	49.8	27.8
32	0.931	1.466	62.4	2344	66.4	57.9	51.6	32.2

Table 6. GRC of the response variable results, their corresponding GRG for each test, and their ranked performance. (All GRC and GRG values are dimensionless quantities ranging from 0 to 1; RSD $Sur_{r,z}$: RSD of surface tension in the radial or axial direction).

Test	GRC								GRG	Rank
	Keyhole Depth	Keyhole Width	RSD Velocity	Mean Temp	RSD $Darcy_r$	RSD $Darcy_z$	RSD Sur_r	RSD Sur_z		
1	0.35	0.45	0.50	0.34	0.55	0.48	0.54	0.88	0.511	15
2	0.45	0.62	0.43	0.45	0.67	0.33	0.41	0.37	0.465	24
3	0.61	0.77	0.39	0.69	0.37	0.41	0.36	0.33	0.491	17
4	1.0	1.00	0.43	1.00	0.34	0.42	0.33	0.34	0.607	5
5	0.39	0.39	0.53	0.34	0.60	0.56	0.65	0.95	0.551	11
6	0.46	0.47	0.48	0.43	0.64	0.41	0.46	0.41	0.470	21
7	0.52	0.56	0.42	0.64	0.40	0.47	0.41	0.34	0.472	20
8	0.69	0.63	0.49	0.88	0.33	0.47	0.39	0.34	0.527	14
9	0.39	0.35	0.51	0.34	0.65	0.86	0.86	0.88	0.605	7
10	0.41	0.40	0.41	0.39	0.53	0.50	0.52	0.57	0.466	23
11	0.46	0.44	0.35	0.51	0.54	0.48	0.44	0.38	0.449	27
12	0.51	0.46	0.37	0.66	0.36	0.51	0.43	0.36	0.457	26
13	0.35	0.33	0.42	0.33	0.71	0.99	1.00	0.62	0.592	8
14	0.34	0.36	0.38	0.34	0.63	0.69	0.64	0.64	0.505	16
15	0.36	0.40	0.41	0.39	0.46	0.45	0.49	0.57	0.440	32
16	0.41	0.46	0.35	0.48	0.46	0.44	0.45	0.49	0.443	31
17	0.33	0.33	0.94	0.33	0.87	0.73	0.83	0.83	0.649	2
18	0.34	0.39	1.00	0.46	0.83	0.61	0.44	0.39	0.559	10
19	0.39	0.43	1.00	0.69	0.38	0.79	0.4	0.41	0.561	9
20	0.45	0.47	0.87	0.98	0.36	0.89	0.38	0.43	0.609	4
21	0.34	0.37	0.57	0.33	0.78	0.95	0.77	0.74	0.607	6
22	0.35	0.44	0.54	0.43	0.72	0.49	0.46	0.42	0.482	19
23	0.41	0.50	0.44	0.63	0.38	0.61	0.40	0.38	0.468	22
24	0.6	0.55	0.42	0.87	0.38	0.65	0.37	0.40	0.530	13
25	0.36	0.36	0.53	0.33	0.86	1.00	0.88	1.00	0.666	1
26	0.42	0.40	0.43	0.38	0.62	0.51	0.52	0.63	0.489	18
27	0.51	0.45	0.35	0.52	0.43	0.48	0.43	0.39	0.444	29
28	0.62	0.49	0.33	0.67	0.37	0.46	0.39	0.37	0.462	25
29	0.38	0.34	0.41	0.34	1.00	0.91	0.94	0.69	0.626	3
30	0.40	0.35	0.38	0.35	0.81	0.65	0.57	0.75	0.532	12
31	0.41	0.38	0.39	0.40	0.53	0.42	0.45	0.60	0.448	28
32	0.45	0.42	0.35	0.48	0.47	0.47	0.44	0.46	0.443	30

Table 7. Results of ANOVA analysis (contribution was calculated as (Adj SS of Factor/total Adj SS) × 100).

Analysis of Variance						
Source	**DOF**	**Adj SS**	**Adj MS**	**F-Value**	***p*-Value**	**Contribution (%)**
Amplitude	3	0.076427	0.025476	111.28	0	51.9
Pulse Shape	1	0.008454	0.008454	36.93	0	5.7
Frequency	3	0.015007	0.005002	21.85	0	10.2
Frequency × Amplitude	9	0.035032	0.003892	17	0	23.8
Pulse Shape × Amplitude	3	0.005416	0.001805	7.89	0.007	3.7
Pulse Shape × Frequency	3	0.004968	0.001656	7.23	0.009	3.4
Error	9	0.00206	0.000229			
Total	31	0.147366				

4.2.1. Keyhole Depth and Width

Analysis of Table 5 shows that the keyhole depth was primarily influenced by the laser power amplitude, with pulse frequency and shape having secondary effects. Increasing the amplitude consistently deepened the keyhole across all profiles. For example, at 100 Hz with sinusoidal pulses, raising the amplitude from 1000 W to 4000 W increased the depth by 54% (0.793 mm to 1.218 mm). A similar trend occurred for other frequencies (Figure 6), though the magnitude of the depth increase diminished at higher frequencies (e.g., 800 Hz). Physically, at lower frequencies, longer pulse durations allowed high-amplitude peaks to deliver sustained energy, resulting in deeper keyholes. In contrast, at 800 Hz, where each pulse lasted approximately 1.25 ms, the shorter duration limited the energy delivered per pulse, reducing the effectiveness of higher peak power on increasing depth. Frequency alone had a mild impact on the depth at low amplitudes but showed a stronger effect at higher amplitudes. For instance, sinusoidal depth dropped by 27% with increasing frequency from 100 Hz to 800 Hz at 4000 W. Cosinusoidal depth also declined but less significantly (5%). This suggests a strong interaction between frequency and amplitude, which is important for optimization. Overall, frequencies from 100 to 400 Hz yielded comparable depths (around 0.9–1.0 mm at mid-to-high amplitudes), with a notable drop at 800 Hz. The slight decline in depth with very high frequency can be attributed to reduced energy per pulse and rapid cycling, which prevented the keyhole from fully developing before the pulse ended. The pulse shape had a subtle yet frequency-dependent influence on penetration. On average, sinusoidal pulses yielded slightly deeper welds (0.928 mm vs. 0.885 mm). At 100 Hz and 4000 W, the sinusoidal profiles produced 24% deeper penetration than the cosinusoidal profiles (1.218 mm vs. 0.984 mm), likely due to their gradual power increase, which facilitated stable keyhole formation before delivering the peak energy, enhancing penetration. Conversely, at 800 Hz, cosinusoidal pulses outperformed the sinusoidal pulses (0.931 mm vs. 0.886 mm), as early power peaks became more effective during brief pulse durations. Overall, frequencies between 100 and 400 Hz at mid-to-high amplitudes provided the most consistent depth performance.

The keyhole width followed trends similar to the depth but showed smaller relative variations. Higher amplitudes slightly increased the width; for example, at 100 Hz, sinusoidal pulses widened the keyhole from 1.50 mm to 1.73 mm (15% increase), compared to a 54% increase in depth. The width generally decreased with higher frequencies. Sinusoidal pulses at 100 Hz produced the widest melt pools (≈1.5 mm at low amplitude), while 800 Hz pulses yielded narrower widths (≈1.35 mm). This trend reflects the effects of longer, more energetic pulses promoting lateral heat diffusion, whereas high-frequency pulsing concentrated energy locally.

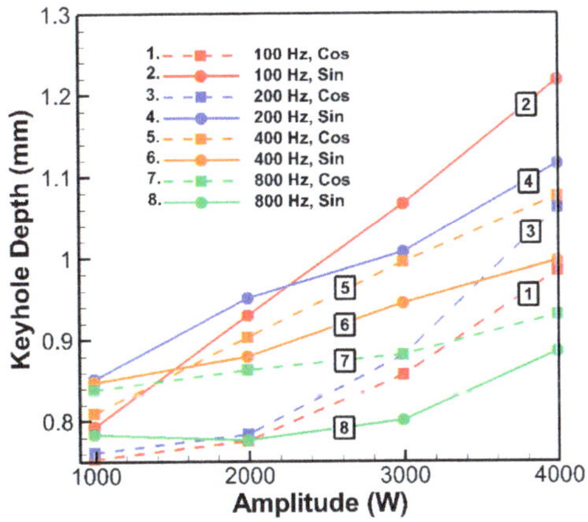

Figure 6. Keyhole depth versus amplitude at different frequencies for sinusoidal and cosinusoidal laser profiles.

4.2.2. Thermal Response (Mean Temperature)

The mean material temperature showed a clear dependence on the pulse parameters. An increasing amplitude consistently reduced the mean temperature across all frequencies. For example, with sinusoidal pulses at 100 Hz, the mean temperature dropped from 2682 K at an amplitude of 1000 W to 1941 K at an amplitude of 4000 W. Similar reductions (≈500–600 K) were observed for cosinusoidal pulses and at other frequencies (Table 5). This inverse relationship is attributed to higher amplitude pulses causing intense but brief heating, followed by longer cooling periods, which lowers the time-averaged temperature. Thus, higher amplitudes not only enhanced penetration but also promoted lower mean temperatures, which may benefit heat-sensitive materials. Conversely, higher frequencies increased the mean temperature by reducing the time between pulses and promoting more continuous energy input. The pulse shape had a negligible direct impact on the mean temperature, which was expected since shape altered the temporal distribution within a pulse but not the overall energy per pulse in these tests.

4.2.3. Process Stability and Fluctuations (Velocity and Forces)

Table 5 reveals that the pulse parameters significantly influenced process stability. Amplitude was the primary contributor to instability, with increased values consistently raising the relative standard deviation (RSD) of key forces. For instance, increasing the amplitude from 1000 W to 4000 W more than doubled the RSD of surface tension force on average (from ~30% to ~57%). Physically, a larger amplitude means higher peak power, causing stronger vapor recoil pressure and fluid flow surges during the pulse peak, followed by a low-power period producing a violent stirring of the melt pool, leading to greater variability. Pulse frequency had a dual effect. Higher frequencies generally reduced the RSDs of surface tension and Darcy damping forces (e.g., averaging over shapes, radial surface tension RSD dropped from 52.7% at 100 Hz to 40.3% at 800 Hz), likely due to more frequent but smaller perturbations. However, the RSD of fluid velocity increased with frequency, from ~41% at 100 Hz to ~58% at 800 Hz (averaging over shapes) as rapid pulsing kept the molten pool in continuous motion, preventing flow stabilization. These frequency effects highlight that different aspects of the process responded differently to the pulse rate: the instantaneous flow became more vigorous with fast repetition, yet the forces driving or resisting that flow might have fluctuated less extremely due to the more continuous application. Pulse shape also played a role in stability, especially at low frequencies. At 100 Hz and 4000 W, cosinusoidal pulses significantly outperformed sinusoidal ones: the velocity RSD was 32.2% vs. 52.8%, the Darcy force RSD (z) was 41.7% vs. 62.2%, and the radial surface tension RSD was 59.7% vs. 67.1%. This may be due to the cosinusoidal profile delivering peak power at the start of the pulse, followed by a gradual decline, allowing oscillations to settle before the next pulse. In contrast, sinusoidal pulses sustained excitation longer, promoting inter-pulse oscillations. However, the stabilizing advantage of the cosinusoidal profiles diminished at higher frequencies. By 400 Hz and 800 Hz, differences between the two shapes became negligible or even reversed slightly. For example, at 800 Hz and 4000 W, the radial surface tension RSD was 51.6% for cosinusoidal and 50.0% for sinusoidal pulses. Consequently, shape–frequency interaction was important in the optimization procedure. Additionally, radial fluctuations were consistently higher than axial ones under the same conditions, indicating that lateral oscillations (widening/narrowing) of the keyhole and melt pool were more pronounced than vertical (depth) fluctuations.

4.3. GRA and ANOVA

To evaluate the overall welding behavior across all eight response variables, the GRA was conducted on the 32 Taguchi-designed trials. Each response was first normalized according to its objective functions specified in Table 4. The GRA analysis and the corresponding ANOVA results are presented in this section.

4.3.1. Response Table for GRG

The GRC for each response (Table 6) was averaged to obtain the GRG for each trial, serving as an overall performance index. Higher GRG values corresponded to deeper and wider welds with reduced fluctuations, instabilities, and thermal load. The GRG values ranged from approximately 0.44 (lowest quality) to 0.67 (highest quality), indicating significant variation in weld quality across different pulsing conditions.

The GRG-based performance ranking reveals clear trends related to pulse shape, frequency, and amplitude. As shown in Table 6, the highest GRG (0.666) was obtained for Test 25 (cosinusoidal, 400 Hz, 1000 W), followed by Test 17 (cosinusoidal, 100 Hz, 1000 W, GRG = 0.649) and Test 29 (cosinusoidal, 800 Hz, 1000 W, GRG = 0.626). Notably, six of the top eight tests used the lowest amplitude (1000 W), indicating that low power amplitude significantly improved the overall weld quality when all eight criteria were considered. Although lower amplitudes resulted in reduced penetration depth, this was offset by substantial improvements in process stability. Stability-related responses (e.g., velocity and force fluctuations) showed GRC values close to 1 (low variability and high stability) at low amplitudes, while high-amplitude conditions often had GRCs near 0.33 (less stability). Conversely, high amplitudes yielded higher GRCs for depth and width, but overall GRG values favored the stability benefits of lower power amplitude. Thus, GRA effectively balanced competing objectives, and the optimal performance occurred at the lowest tested amplitude. Pulse frequency also influenced weld quality. The average GRG was highest at 100 Hz (~0.557) and declined with increasing frequency. However, the top-ranked case occurred at 400 Hz, suggesting that an optimal combination of pulse shape and amplitude can compensate for the negative effects of higher frequency. This highlights the importance of interaction effects between parameters. Pulse shape had a smaller main effect compared to amplitude and frequency, but still played a role. On average, the cosinusoidal profile yielded a higher mean GRG (0.536) than the sinusoidal profile (0.504). This was especially evident at low amplitude, where early peak power in cosinusoidal pulses likely aided initial keyhole formation without destabilizing the melt pool. For example, at 100 Hz and 1000 W, the cosinusoidal case (Test 17, GRG = 0.649) significantly outperformed its sinusoidal counterpart (Test 1, GRG = 0.512), suggesting that an early power surge was beneficial at low frequency. Similar trends were observed at 400 Hz and 800 Hz. However, at higher amplitudes or when averaged across all frequencies, the shape advantage diminished. Thus, pulse shape alone had a modest effect (ranked third among the three factors), but its impact emerged in combination with frequency and amplitude settings. Overall, amplitude had the strongest influence on GRG, followed by frequency, and then shape. Lower amplitude and frequency, especially with a cosinusoidal waveform, provided the most favorable combination of weld quality and process stability. Conversely, the lowest-ranked tests often involved high frequencies and higher amplitudes with sinusoidal profiles.

4.3.2. ANOVA Analysis of GRG and Key Factor Effects

To evaluate the statistical significance and influence of each control factor on overall welding performance, ANOVA was conducted on the GRG results. Table 7 summarizes the effects of the three main factors, including pulse shape (A), frequency (B), and amplitude (C), and their two-way interactions (A × B, A × C, B × C). The analysis revealed that all three main factors and the major interactions had a significant effect on the GRG (p-values < 0.01). The error term was very small (only ~1.4% of the total variation), indicating that the chosen factors and interactions explained ~98.6% of the variability in the multi-response performance. This confirms the effectiveness of the proposed optimization design in capturing the key influences on weld quality.

Among the main factors, laser power amplitude (Factor C) had the strongest influence, accounting for 51.9% of the total GRG variation. This quantitatively confirms its dominant role, as previously observed in the GRA trends. While higher amplitudes improved penetration, they also increased melt pool turbulence, reducing the overall GRG. In contrast, lower amplitudes enhanced stability and yielded a better balance between penetration and process control, supporting the recommendation to minimize amplitude within practical limits for improved weld quality. Pulse frequency was the third most influential factor (10.2% of GRG variation), followed by pulse shape (5.7%), which also showed statistically significant effects. Cosinusoidal waveforms generally outperformed sinusoidal ones, especially at low amplitudes and frequencies, due to their early energy delivery, which promoted more stable keyhole formation. Significant interaction effects were also observed: frequency–amplitude accounted for 23.8% of GRG variation, while shape–frequency and shape–amplitude interactions contributed approximately 3–4% each. These results highlight the importance of tuning parameter combinations rather than individual factors alone. For instance, high amplitude was more tolerable at low frequencies but led to substantial instability at high frequencies. Similarly, the benefits of cosinusoidal pulses were most pronounced under low-frequency, low-amplitude conditions, diminishing at higher frequencies.

5. Conclusions

This study introduced a Taguchi-based multi-response optimization approach using GRA to determine the optimal conditions for keyhole-mode laser spot welding of aluminum. The effects of a novel cosinusoidal laser power profile were evaluated alongside sinusoidal and continuous profiles. Welding performance was analyzed and optimized based on eight key response variables, including keyhole depth and width (larger-the-better), and the relative standard deviations of surface tension, Darcy damping force, and fluid velocity (smaller-the-better). The key findings are summarized as follows:

- Seven of the first ten best cases in terms of the overall welding performance belonged to the cosinusoidal pulse shape. A cosinusoidal pulse shape (test 25) with 1000 W amplitude and 400 Hz was found to be the best case, while a sinusoidal counterpart (test 15) with 800 Hz and 4000 W amplitude had the worst performance.

- The GRA results demonstrate that the best welding performances were achieved at a low amplitude of 1000 W (six among the top ten performances). However, increasing the amplitude to its maximum (4000 W) reached its best performance (fourth place among all) only when the frequency was set to its minimum (100 Hz).
- High-frequency and high-power amplitude pulses tend to destabilize the process and degrade the multi-objective outcome, whereas low-amplitude power, slow pulses foster stable keyhole dynamics and uniform heating, even if the penetration is lower.
- Selecting lower frequencies (100–400 Hz) is crucial for maximizing weld quality, offering adequate penetration with significantly improved process stability.
- Amplitude, frequency, and the two-way interaction between them contributed the most to the weld quality and system performance, with amplitude having the most contribution.

Author Contributions: S.S.: Methodology, software, validation, conceptualization, visualization, investigation, formal analysis, writing—original draft, and writing—review and editing. X.-T.P.: Methodology, software, validation, conceptualization, visualization, investigation, formal analysis, writing—original draft, supervision, and writing—review and editing. All authors have read and agreed to the published version of the manuscript.

Funding: This research was funded by the Natural Sciences and Engineering Research Council of Canada (NSERC), grant number Discovery Grant (RGPIN-2019-06574); and Fonds de recherche du Québec, Nature et Technologies (FRQNT), grant number 2024-2025-B2X-343090.

Institutional Review Board Statement: Not applicable.

Informed Consent Statement: Not applicable.

Data Availability Statement: The original contributions presented in the study are included in the article. Further inquiries can be directed to the corresponding author.

Acknowledgments: The authors would like to acknowledge the valuable efforts and time of the anonymous reviewers who took the time to help improve the quality of this paper during the revision process.

Conflicts of Interest: The authors declare no conflicts of interest. The funding bodies had no role in the design of this study; in the collection, analysis, or interpretation of data; in the writing of the manuscript, or in the decision to publish the results.

Nomenclatures

T_m	Melting temperature; [K]
T_V	Vaporization temperature; [K]
T_s	Solidus temperature; [K]
T	Temperature; [K]
dT_m	Smoothing interval of melting; [K]
dT_V	Smoothing interval of vaporization; [K]
k_s	Thermal conductivity of solid; [W/m/K]
k_l	Thermal conductivity of liquid; [W/m/K]
L_m	Latent heat of fusion; [J/kg]
L_v	Latent heat of evaporation; [J/kg]
R	Universal gas constant; [J/mol/K]
Cp_s	Specific heat of solid; [J/kg/K]
Cp_l	Specific heat of liquid; [J/kg/K]
μ_s	Dynamic viscosity of solid; [Pa.s]
μ_l	Dynamic viscosity of liquid; [Pa.s]

d	Form factor for Gaussian distribution
C	Coefficient in Darcy's law
b	Coefficient in Darcy's law
R_{eff}	Effective radius of a laser beam; [m]
d	Dendrite dimension; [m]
M	Molecular mass of aluminum; [kg/mol]
h	Convective heat transfer coefficient; [$W/m^2/K$]
f	Laser frequency; [Hz]
A	Laser power amplitude [W]
$\vec{g}$	Gravity; [m/s^2]
p	Pressure; [atm]
$\vec{u}$	Velocity; [m/s]
t	Time; [s]
F_{Darcy}	Darcy damping force; [N/m^3]
$F_{Buoyancy}$	Buoyancy force; [N/m^3]
$V_{f,1}$	Volume fraction of fluid 1
$V_{f,2}$	Volume fraction of fluid 2
D_m	Gauss function around the melting temperature
D_V	Gauss function around the vaporization temperature
K	Constant representing the mushy zone morphology; [$1/m^2$]
P_{sat}	Saturated vapor pressure; [atm]
P_{atm}	Atmospheric pressure; [atm]
V_L	Volume fraction of liquid
V_s	Volume fraction of solid
$\vec{n}$	Normal vector on the vapor/liquid interface
$\vec{k}$	Tangential vector on the vapor/liquid interface
B_t	Temporal laser distribution function used to apply pulses
n	Simulation steps
$x_j^{(i)}(t)$	Time-dependent value of the j-th response for the i-th test at time step t
$\overline{x}_j^{(i)}$	Mean of $x_j^{(i)}(t)$ over all simulation steps sampled from t = 0 to t = 0.01 s
X_{ij}	Original values of the j-th response of the i-th test
$X_{ij}{}^*$	Normalized values of the j-th response of the i-th test
X_j	Set of all values of the j-th response across all test cases ($X_j = \left\{X_{1j}, X_{2j}, \ldots X_{32j}\right\}$).
Δ_{ij}	Deviation of the normalized response from the ideal normalized response
Δ_{min}	Minimum possible deviation
Δ_{max}	Maximum possible deviation
r	Radial direction in the 2D axisymmetric design
z	Axial direction in the 2D axisymmetric design
Greek	
γ	Level-set parameter; [m/s]
Δ	Deviation between two values
ε	Level-set parameter; [m]
δ	Delta function
ϕ	Level-set function (variable)
α	Absorptivity of aluminum on 1064 nm laser
ξ	Surface emissivity; distinguishing coefficient ranges from 0 to 1
β_L	Thermal expansion coefficient; [1/K]
β_R	Retro-diffusion coefficient
ρ	Density; [kg/m^3]
μ	Dynamic viscosity; [Pa.s]
σ	Surface tension coefficient; [N/m]
Subscript	
L	Liquid

V	Vapor/vaporization
m	Melting; total number of responses in the GRA approach
Vol	Volume force
g	Gas
st	Surface tension
ls	Level set
Abbreviation	
ANOVA	Analysis of variance
GRA	Grey relational analysis
GRC	Grey relational coefficient
GRG	Grey relational grade
RSD	Relative standard deviation
SD	Standard deviation
LS	Level set
TEPT	Thermal-enthalpy porosity technique
MMT	Modified mixture theory
PW	Pulsed wave
CW	Continuous wave
PWM	Power wave modulation
HAZ	Heat affected zone
DOF	Degrees of freedom
OA	Orthogonal array
Sur	Surface tension
Adj	Adjusted
SS	Sum of squares
MS	Mean squares
CV	Cross-sectional view
TV	Top view

References

1. Miller, W.S.; Zhuang, L.; Bottema, J.; Wittebrood, A.J.; De Smet, P.; Haszler, A.; Vieregge, A. Recent development in aluminium alloys for the automotive industry. *Mater. Sci. Eng. A* **2000**, *280*, 37–49. [CrossRef]
2. Bunaziv, I.; Hovig, E.W.; Godinez Brizuela, O.E.; Zhang, K.; Ma, X.; Ren, X.; Eriksson, M.; Skjetne, P. CFD modeling for predicting imperfections in laser welding and additive manufacturing of aluminum alloys. *J. Laser Appl.* **2024**, *36*, 032010. [CrossRef]
3. Schubert, E.; Klassen, M.; Zerner, I.; Walz, C.; Sepold, G. Light-weight structures produced by laser beam joining for future applications in automobile and aerospace industry. *J. Mater. Process. Technol.* **2001**, *115*, 2–8. [CrossRef]
4. Wang, L.; Yao, M.; Gao, X.; Kong, F.; Tang, J.; Jun Kim, M. Keyhole stability and surface quality during novel adjustable-ring mode laser (ARM) ssswelding of aluminum alloy. *Opt. Laser Technol.* **2023**, *161*, 109202. [CrossRef]
5. Bunaziv, I.; Akselsen, O.M.; Ren, X.; Nyhus, B.; Eriksson, M. Laser Beam and Laser-Arc Hybrid Welding of Aluminium Alloys. *Metals* **2021**, *11*, 1150. [CrossRef]
6. Cavilha Neto, F.; Pereira, M.; dos Santos Paes, L.E.; Fredel, M.C. Assessment of power modulation formats on penetration depth for laser welding. *J. Braz. Soc. Mech. Sci. Eng.* **2021**, *43*, 286. [CrossRef]
7. Hajavifard, R.; Motahari, M.; Özden, H.; Miyanaji, H.; Kafashi, S. The Effects of Pulse Shaping Variation in Laser Spot-Welding of Aluminum. *Procedia Manuf.* **2016**, *5*, 232–247. [CrossRef]
8. Svenungsson, J.; Choquet, I.; Kaplan, A.F.H. Laser Welding Process—A Review of Keyhole Welding Modelling. *Phys. Procedia* **2015**, *78*, 182–191. [CrossRef]
9. Fetzer, F.; Sommer, M.; Weber, R.; Weberpals, J.-P.; Graf, T. Reduction of pores by means of laser beam oscillation during remote welding of AlMgSi. *Opt. Lasers Eng.* **2018**, *108*, 68–77. [CrossRef]
10. Pang, S.; Chen, W.; Zhou, J.; Liao, D. Self-consistent modeling of keyhole and weld pool dynamics in tandem dual beam laser welding of aluminum alloy. *J. Mater. Process. Technol.* **2015**, *217*, 131–143. [CrossRef]
11. Chen, G.; Liu, J.; Shu, X.; Gu, H.; Zhang, B. Numerical simulation of keyhole morphology and molten pool flow behavior in aluminum alloy electron-beam welding. *Int. J. Heat Mass Transf.* **2019**, *138*, 879–888. [CrossRef]
12. SaediArdahaei, S.; Pham, X.-T. Toward Stabilizing the Keyhole in Laser Spot Welding of Aluminum: Numerical Analysis. *Materials* **2024**, *17*, 4741. [CrossRef] [PubMed]

13. SaediArdahaei, S.; Pham, X.-T. Comparative Numerical Analysis of Keyhole Shape and Penetration Depth in Laser Spot Welding of Aluminum with Power Wave Modulation. *Thermo* **2024**, *4*, 222–251. [CrossRef]
14. Matsunawa, A.; Kim, J.-D.; Seto, N.; Mizutani, M.; Katayama, S. Dynamics of keyhole and molten pool in laser welding. *J. Laser Appl.* **1998**, *10*, 247–254. [CrossRef]
15. Tsukamoto, S.; Kawaguchi, I.; Arakane, G.; Honda, H. Keyhole Behaviour in High Power Laser Welding. In *First International Symposium on High-Power Laser Macroprocessing*; SPIE: Bellingham, DC, USA, 2003; pp. 251–256.
16. Zhang, M.; Wu, J.; Mao, C.; Cheng, B.; Shakhawat, H.M.D.; Li, H.; Wang, K.; Zhang, J.; Hu, Y.; Bi, Z. Impact of power modulation on weld appearance and mechanical properties during laser welding of AZ31B magnesium alloy. *Opt. Laser Technol.* **2022**, *156*, 108490. [CrossRef]
17. Heider, A.; Weber, R.; Herrmann, D.; Herzog, P.; Graf, T. Power modulation to stabilize laser welding of copper. *J. Laser Appl.* **2015**, *27*, 022003. [CrossRef]
18. Omoniyi, P.; Acharya, U.; Akinlabi, S.; Jen, T.-C.; Akinlabi, E. Taguchi-Grey relational analysis driven multi-objective optimization of weld bead geometry and hardness in laser welded Ti6Al4V Alloy. *Int. J. Interact. Des. Manuf. IJIDeM* **2025**, *19*, 423–434. [CrossRef]
19. Tsai, M.-J.; Wu, L.-F. Multi-Objective Optimization of Nd: YAG Laser Drilling of Optical-Grade Acrylic Plate Using Taguchi-Based Grey Relational Analysis. *Materials* **2022**, *15*, 8998. [CrossRef]
20. Cai, J.; Jin, H.; Ouyang, Z.; Guo, K.; Wei, Y.; Chen, J. Study on asymmetric forming and performance strengthening mechanisms of 6061-T6 aluminum alloy joints fabricated by oscillating laser based on keyhole characteristics and molten pool dynamic behaviors. *Int. J. Therm. Sci.* **2025**, *207*, 109377. [CrossRef]
21. Wen, X.; Wu, D.; Zhang, P.; Liu, S.; Luo, Z.; Jia, Z.; Ye, X.; Shi, H.; Yu, Z. Influence mechanism of the keyhole behavior on penetration depth by in-situ monitoring in pulsed laser welding of aluminum alloy. *Optik* **2021**, *246*, 167812. [CrossRef]
22. Olsson, E.; Kreiss, G. A conservative level set method for two phase flow. *J. Comput. Phys.* **2005**, *210*, 225–246. [CrossRef]
23. Kong, F.; Kovacevic, R. Modeling of Heat Transfer and Fluid Flow in the Laser Multilayered Cladding Process. *Metall. Mater. Trans. B* **2010**, *41*, 1310–1320. [CrossRef]
24. Ardahaie, S.S.; Hosseini, M.J.; Ranjbar, A.A.; Rahimi, M. Energy storage in latent heat storage of a solar thermal system using a novel flat spiral tube heat exchanger. *Appl. Therm. Eng.* **2019**, *159*, 113900. [CrossRef]
25. Zhang, Y.; Shen, Z.; Ni, X. Modeling and simulation on long pulse laser drilling processing. *Int. J. Heat Mass Transf.* **2014**, *73*, 429–437. [CrossRef]
26. Mayi, Y.A.; Dal, M.; Peyre, P.; Bellet, M.; Metton, C.; Moriconi, C.; Fabbro, R. Laser-induced plume investigated by finite element modelling and scaling of particle entrainment in laser powder bed fusion. *J. Phys. Appl. Phys.* **2020**, *53*, 075306. [CrossRef]
27. Courtois, M.; Carin, M.; Le Masson, P.; Gaied, S.; Balabane, M. A complete model of keyhole and melt pool dynamics to analyze instabilities and collapse during laser welding. *J. Laser Appl.* **2014**, *26*, 042001. [CrossRef]
28. Pang, S.; Chen, L.; Zhou, J.; Yin, Y.; Chen, T. A three-dimensional sharp interface model for self-consistent keyhole and weld pool dynamics in deep penetration laser welding. *J. Phys. Appl. Phys.* **2011**, *44*, 025301. [CrossRef]
29. Pahamli, Y.; Hosseini, M.J.; Ardahaie, S.S.; Ranjbar, A.A. Improvement of a phase change heat storage system by Blossom-Shaped Fins: Energy analysis. *Renew. Energy* **2022**, *182*, 192–215. [CrossRef]
30. Ardahaie, S.S.; Hosseini, M.J.; Eisapour, M.; Eisapour, A.H.; Ranjbar, A.A. A novel porous metal hydride tank for hydrogen energy storage and consumption assisted by PCM jackets and spiral tubes. *J. Clean. Prod.* **2021**, *311*, 127674. [CrossRef]
31. Pakrouh, R.; Hosseini, M.J.; Ranjbar, A.A.; Bahrampoury, R. A numerical method for PCM-based pin fin heat sinks optimization. *Energy Convers. Manag.* **2015**, *103*, 542–552. [CrossRef]
32. Kuo, Y.; Yang, T.; Huang, G.-W. The use of grey relational analysis in solving multiple attribute decision-making problems. *Comput. Ind. Eng.* **2008**, *55*, 80–93. [CrossRef]

materials

Article

Influence of Surface Damage on Weld Quality and Joint Strength of Collision-Welded Aluminium Joints

Stefan Oliver Kraus, Johannes Bruder, Florian Schuller and Peter Groche *

Institute for Production Engineering and Forming Machines, TU Darmstadt, D-64287 Darmstadt, Germany; stefan.kraus@ptu.tu-darmstadt.de (S.O.K.); johannes.bruder@ptu.tu-darmstadt.de (J.B.)
* Correspondence: peter.groche@ptu.tu-darmstadt.de

Abstract

Collision welding represents a promising solid-state joining technique for combining both similar and dissimilar metals without the thermal degradation of mechanical properties typically associated with fusion-based methods. This makes it particularly attractive for lightweight structural applications. In the context of collision welding, it is typically assumed that ideally smooth and defect-free surface conditions exist prior to welding. However, this does not consistently reflect industrial realities, where surface imperfections such as scratches are often unavoidable. Despite this, the influence of such surface irregularities on weld integrity and quality has not been comprehensively investigated to date. In this study, collision welding is applied to the material combination of AA6110A-T6 and AA6060-T6. Initially, the process window for this material combination is determined by systematically varying the collision velocity and collision angle—the two primary process parameters—using a special model test rig. Subsequently, the effect of surface imperfections in the form of defined scratch geometries on the resulting weld quality is investigated. In addition to evaluating the welding ratio and tensile shear strength, weld quality is assessed through scanning electron microscopy (SEM) of the bonding interface and high-speed imaging of jet formation during the collision process.

Keywords: collision welding; impact welding; electromagnetic pulse welding; welding window; metallographic weld zone; multi-material design; lightweight concepts; weight reduction; emission reduction

1. Introduction

Mobility is a fundamental aspect of modern society, enabling access to education, healthcare, and the labour market. It includes private and public transport, logistics, and infrastructure [1]. In 2023, the transport sector accounted for 21% of global CO_2 emissions, 38% of which came from passenger cars [2–4]. In addition to greenhouse gases, non-exhaust particulate emissions–especially from tyre wear and brake dust–are gaining attention as they are strongly influenced by vehicle weight [5–13]. Reducing vehicle mass, therefore, helps to reduce both CO_2 emissions and particulate pollution such as microplastics [14–16]. This underlines the environmental importance of lightweight design [17].

These lightweighting goals can be achieved by using high-strength aluminium alloys, which are commonly used in automotive engineering [18]. However, joining these alloys using conventional fusion-based welding techniques presents significant challenges. These include the formation of a heat-affected zone (HAZ) leading to strength degradation, porosity caused by stable oxide layers, distortion due to the high thermal conductivity of aluminium, and hot cracking associated with alloying elements [19–22]. To exploit the lightweight potential of high-strength aluminium alloys, a joining method is required that mitigates these challenges, most of which are directly attributable to the thermal input inherent in conventional welding processes.

In this area, the solid state welding or collision welding process group opens up new production potential. The joining mechanism is based on the application of sufficiently high pressure to join similar and dissimilar material combinations without active heat input. Therefore, the formation of intermetallic phases in the joint zone is avoided [23–25]. Joining feasibility is defined by collision velocity and angle, with the process window categorised into solid-phase, liquid-phase, and hybrid joining regions [26]. Electromagnetic pulse welding (EMPW) facilitates collision welding in industrial settings by accelerating a joining partner via a strong electromagnetic field, creating a high-speed impact weld [25,27–29].

Due to transient effects, both the collision velocity and the collision angle change dynamically during the EMPW collision process, directly influencing the properties of the resulting joint [30,31]. Currently, EMPW process design is carried out exclusively in an iterative manner, which involves considerable time and cost expenditure and leads to uncertainties regarding both joint performance and process stability [32]. The existing iterative process design is based on idealised "optimal" conditions with cleaned and undamaged material surfaces. However, such conditions do not always exist in industrial environments.

Fine scratches on component surfaces can result from upstream manufacturing processes such as rolling or extrusion. In addition, handling, transport, and storage of semi-finished products are particularly relevant as potential sources of surface damage in the form of deeper scratches [33]. The correlation between the sensitivity of the collision welding process to environmental influences-demonstrated in previous studies, for example, through varying surface roughness [34]. The occurrence of scratches in the production process suggests that such scratches, under otherwise identical process parameters, may lead to fluctuating joint strengths or even to a complete failure of the weld. Such a condition is undesirable from a quality assurance perspective in industrial manufacturing processes and is unacceptable due to the potentially high degree of process uncertainty.

There are two basic approaches to solving this problem, both of which require an assessment of the component surfaces.

1. Use of exclusively undamaged/scratch-free components:
   This approach requires considerable effort in terms of handling and storage of the semi-finished products within the production process. In addition, it leads to high rejection rates, as even components with minor damage from previous process steps must be sorted out.
2. Identification and evaluation of critical scratches:
   In this approach, only components with scratches deemed critical are discarded, or alternatively, the process parameters—such as collision velocity—are adjusted so that a sufficiently strong joint is achieved even in the presence of non-critical scratches.

The second approach significantly reduces the rejection of components and semi-finished products and allows for less complex concepts for component handling and storage, as only the exceeding of critical scratch characteristics needs to be prevented. This contributes significantly to cost reduction and resource savings, as the need for semi-finished products, packaging protection materials, and complex control measures is reduced.

In the context of this study, a model test rig for collision welding is used to investigate the influence of scratches on joint quality systematically. In this process, scratch depth, number, and orientation relative to the welding direction are varied. The aim is to identify critical thresholds that will define the minimum surface quality requirements that must not be undershot by manufacturing, handling, and storage processes. This enables the targeted selection of suitable process routes in an industrial setting with minimal effort. The experimental setup, as well as the tests and analyses conducted, are detailed in the following section.

2. Materials and Methods

The experimental procedure is divided into three process steps. First, the specimens are prepared. Subsequently, they are subjected to collision welding under defined collision angles and collision velocities using the model test rig. After the collision welding tests, the specimens are analysed using various measurement techniques to evaluate the joint formation and the resulting weld quality.

The specimens to be welded (collision area: length × width: 12.5 mm × 12 mm) are taken from AA6060-T6 extruded flat material (flyer, thickness: 2.0 mm, tensile strength R_m: 215 N/mm^2) and AA6110A-T6 extruded flat material (target, thickness: 2.0 mm, tensile strength R_m: 290 N/mm^2) by shear cutting. The materials' chemical composition corresponds to DIN EN 573-3 [35]. The proportions of the alloying elements are listed in Table 1. For each material, three tensile tests are performed on a Zwick Roell 100 combined tensile (Ulm, Germany) and compression testing machine to determine tensile strength in accordance with DIN 50125 [36]. The samples are produced so that the extrusion direction of the material is parallel to the welding direction.

Table 1. The chemical composition of the aluminium alloys in accordance with DIN EN 573-3.

Alloy	Alloying Elements								
	Si	Fe	Cu	Mn	Mg	Cr	Zn	Ti	Others
AA6060-T6	0.30–0.60	0.10–0.30	0.10	0.10	0.35–0.60	0.05	0.15	0.10	0.15
AA6110A-T6	0.70–1.10	0.50	0.30–0.80	0.30–0.90	0.70–1.10	0.05–0.25	0.20	0.20	0.15

The sample preparation is carried out depending on the specific experimental configuration. Smooth, undamaged samples are cleaned with acetone before the test and visually inspected for surface defects. If the influence of scratches is to be specifically investigated, these are manually introduced into the samples. The introduction of the defined scratches is carried out using the device shown in Figure 1a. In this case, a spring-loaded, hardened needle with a tip angle of 30° is guided over the sample surface. The device allows for precise adjustment of the needle protrusion and scratch depth.

Through flexible positioning of the samples in special fixtures in the device, it is possible to rotate the samples so that scratches can be applied parallel, perpendicular, or at an angle (e.g., 45° rotation) to the welding direction. Additionally, the samples can be offset laterally to allow multiple parallel scratches to be introduced on the sample surface at a distance of 4 mm from each other. In this study, the influence of a scratch parallel to the welding direction and one or two scratches perpendicular to the welding direction is investigated, as shown in Figure 1c. The scratches are introduced into the surface of the AA6060-T6 (flyer) in the used aluminium-aluminium material combination, as this alloy, due to its lower strength, experiences scratches more quickly in industrial production environments compared to the harder AA6110A-T6. Due to its lower strength, AA6060-T6 is used as the flyer in the experiments, as this reduces the stress on the flyer rotor when the predetermined breaking point is torn (see [37]). Furthermore, the selection of the lower-strength material as the flyer allows for better transferability to future scratch investigations with multi-material combinations (e.g., aluminium-steel, and the EMPW process). On the one hand, when using a combination of aluminium and steel in the EMPW process, aluminium is always used as the flyer material. This is due to its much higher electrical conductivity compared to steel [38]. On the other hand, aluminium will always show a higher sensitivity to scratches.

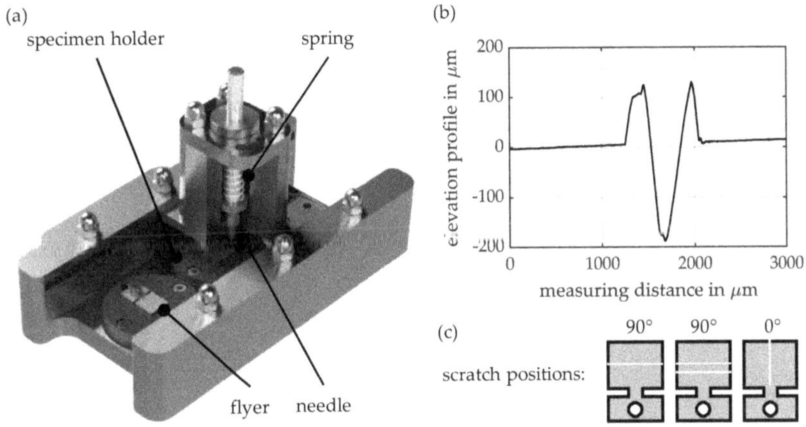

Figure 1. (**a**) The experimental configuration of the scratching device for perpendicular scratches. (**b**) Height profile of a scratch on a prepared specimen (scratch depth 3). (**c**) Representation of the scratch designs to be investigated.

The determination of the scratch depth is carried out using the tactile roughness measurement device Hommel Etamic-T 8000 (Villingen-Schwenningen, Germany). The height profile of the prepared samples is recorded with two equidistant measurement lines perpendicular to the direction of the scratch. The scratch depth is then extracted from the height profile and averaged over the two measurements. Figure 1b shows an example of the height profile of a sample prepared by scratching to the greatest examined depth at a 90° orientation. The scratch of the specific sample has an indented depth of 190 μm. The needle scratches the sample surface, displacing material and forming elevations on both sides of the scratch. In the sample shown, these elevations are 125 μm high. The ratio of the height of the side elevations to the actual scratch depth remains almost constant across different scratch depths. In the following sections, scratch depth is defined as the depth of the scratch in relation to the zero line.

Table 2 summarises the scratch parameters investigated in this study and their corresponding designations. For each parameter combination, three specimens are welded and examined. The combinations are differentiated based on the number of scratches per specimen (scr), scratch orientation (0°: parallel to the welding direction; 90°: perpendicular to the welding direction), and scratch depth (d). The scratch depth is divided into three ranges: d1 (15–30 μm), d2 (50–80 μm), and d3 (170–200 μm). The division into ranges is made because, due to the fluctuating material thickness (±0.05 mm) and the necessary clearance of the scratch device, there is always some variation in the introduced scratch depth. The comparatively large scratch depths are selected to investigate potential surface damage that can occur during industrial processes, such as vibration feeders or dropping materials into transport boxes without protection. The wide range of depths is used to determine the maximum depth up to which welding with good strength values is possible.

Table 2. Summarised scratch parameters.

Description	Number of Scratches (scr)	Orientation	Depth (d) in µm
1scr-0°-d1	1	0°	15–30
1scr-0°-d2	1	0°	50–80
1scr-0°-d3	1	0°	170–200
1scr-90°-d1	1	90°	15–30
1scr-90°-d2	1	90°	50–80
1scr-90°-d3	1	90°	170–200
2scr-90°-d1	2	90°	15–30
2scr-90°-d2	2	90°	50–80
2scr-90°-d3	2	90°	170–200

The collision welding process is carried out on the institute's model test rig for collision welding, which is shown in Figure 2a. Two specimens—the so-called flyer and the so-called target—are accelerated to a defined collision velocity v_{imp} using rotating drives and brought into collision at a predetermined collision angle β. The collision angle is set by bending the flyer, while the collision velocity is adjusted through the motor speeds. A detailed description of the test rig can be found in source [37].

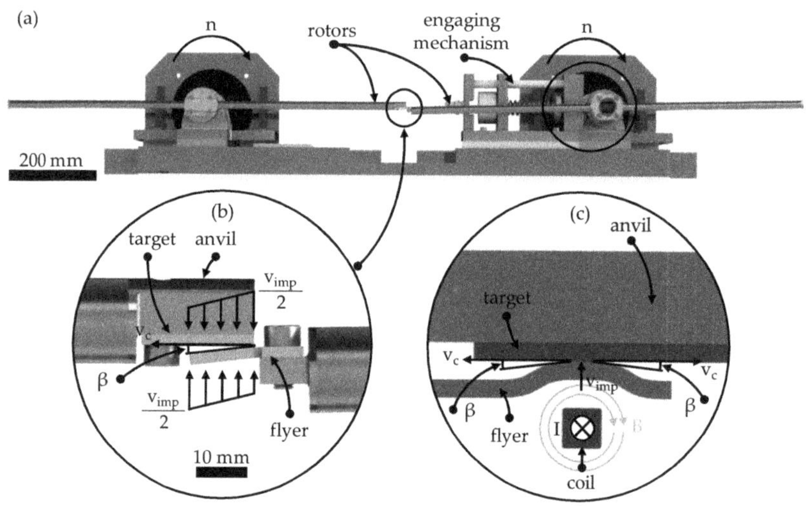

Figure 2. (**a**) The model test rig consists of two synchronised rotating rotors and an engaging mechanism. (**b**) The specimens are mounted at the end of each rotor. (**c**) Process setup EMPW according to [39].

The detailed view (Figure 2b) shows the two specimens immediately before impact. Additionally, the characteristic process parameters—collision velocity v_{imp}, collision point velocity v_c, and collision angle β—are highlighted. The collision point velocity can be calculated using Equation (1), where the collision velocity v_{imp} is the velocity of the specimen perpendicular to the flyer [26].

$$\tan\beta = \frac{v_{imp}}{v_c} \quad (1)$$

During the collision, a rolling motion of the flyer occurs with the collision point velocity v_c on the rear-supported target, which leads to joint formation between the two joining partners through the formation of a jet. This process forms the basis of joint formation in all collision welding techniques, including the previously mentioned EMPW. In contrast to EMPW with transient collision conditions (where collision velocity and angle vary over the course of the collision, see Figure 2c, the main feature of the test rig is the ability to precisely adjust and maintain constant process parameters over time, as well as the good observability of the process.

By systematically varying the collision angle and velocity, a characteristic process window for the material pairing is determined, which provides insight into the parameter space in which a solid-state connection is possible. The collision velocity v_{imp} is varied in six steps: 279 m/s, 297 m/s, 305 m/s, 314 m/s, 331 m/s, and 349 m/s. High-speed optical observation is performed using an HSFC Pro image intensifier camera from PCO (Kelheim, Germany) and a Milvus Macro 100 mm f2.0 macro lens from Zeiss (Oberkochen, Germany). The camera allows exposure times of <20 ns and the acquisition of up to eight images per collision. The brightness of the high-speed images is ensured by a CAVILUX Smart illumination laser from Cavitar (power: 400 W, wavelength: 640 nm, Tampere, Finland).

The procedure for the subsequent sample analysis is shown in Figure 3 and is divided into three consecutive steps.

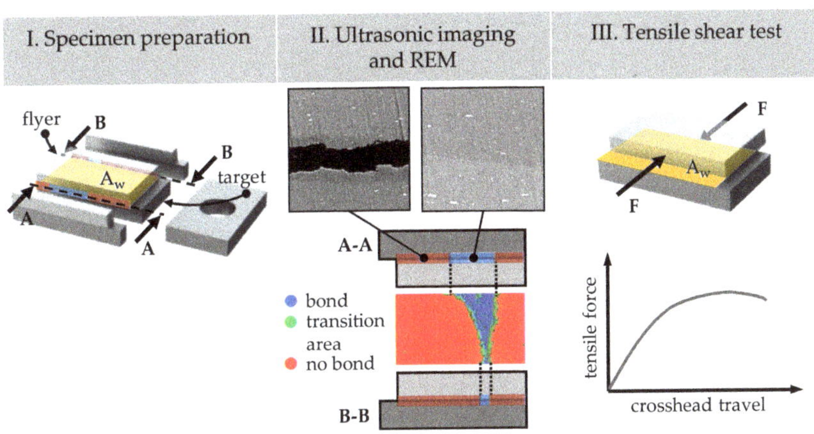

Figure 3. Scheme of the subsequent specimen analysis procedure.

I. Sample Preparation

For the analysis of the welding zone, the edge areas and the mounting surface of the welded samples are first cut off using a wet-cutting machine. The purpose of this is to remove any possible manufacturing influences in the edge areas due to the specimen preparation and to ensure a defined and fully overlapping joining surface between the flyer and the target. The reason for this is that due to process tolerances of the test rig, a precise 100% overlap between the flyer and target cannot be guaranteed. By trimming to a width of 8 mm ± 1 mm and a length of 14 mm ± 1 mm, a standardised/comparable sample geometry is ensured. Subsequently, the lateral separation surfaces of the samples are planed on both sides with SiC paper (grit sizes P320, P400, P600, P800, P1200, P2500, and P4000) and polished with diamond suspension (3 μm and 1 μm) to enable precise analysis of the joining zone.

II. Ultrasonic examinations and SEM

The welding ratio is determined using an ultrasonic measurement system of the type MiniScanner from the manufacturer Amsterdam Technology (Zwinderen, The Netherlands). Its probe captures a measurement area in the form of a long slot, measuring 25.0 mm in length and 11.4 mm in width, with a combined rotational and translational movement.

The scanner's recording provides a graphical representation of the welded (blue) and non-welded (red) areas, as well as the transition zones (green) within the samples. The welding ratio, the ratio of the welded area to the overlapping area A_W, is analysed using a Matlab script. Therefore, the transition area is also divided into welded and non-welded areas using stored threshold values for the ultrasonic signals. The accuracy and reliability of the ultrasonic imaging results in this study are validated by comparing them with metallographic cross-sectional images. The exact nature of the transition zone is still not fully understood and remains a subject of further investigation. The microscopic imaging of the cross-sectional images of the weld zone is carried out using a Phenom ProX Scanning Electron Microscope (SEM) manufactured by Phenom-World (Waltham, MA, USA), as well as an MIRA3 SEM from TESCAN (Brno, Czech Republic).

III. Tensile Shear Test

To determine the shear tensile strength of the welded specimens, they are clamped into a specially designed test setup and then loaded on a Zwick Roell 100 combined tensile and compression testing machine at a test speed of 0.1 mm/s until separation occurs. A detailed description of the test setup and the operation of the device can be found in source [26]. The shear tensile strength is calculated from the previously measured overlapping bonding area A_W and the measured tensile force. This enables a comparable assessment of the mechanical joint quality. The repeatability of the tensile shear tests depends on the overall Zwick Roell 100 combined tensile and compression testing machine with an integrated force sensor.

3. Results

3.1. Welding Process Window

Figure 4 shows the experimentally determined process window for the investigated material combination of AA6060-T6 (thickness 2.0 mm) and AA6110A-T6 (thickness 2.0 mm) in the velocity range up to a maximum of 349 m/s and collision angles between 2.0° and 9.5°. The first joints are observed at a collision velocity of 279 m/s and a collision angle of 4.8°. A collision welding test is classified as a joined test if manual separation of the specimens is not possible after the test (black dots). To improve the clarity of the diagram, the unsuccessful tests are shifted 1 m/s to the right of the successful tests on the collision velocity axis (white dots).

As the collision velocity increases, the angular range within which a connection is formed between the joining partners increases. A key parameter for describing this range is the so-called limit angle, which describes the minimum or maximum collision angle at which a joint is still possible at a given velocity. The observed increase in the range in which joints occur with increasing velocity has been documented several times in the literature, see [40,41]. The behaviour of the lower limit angle varies depending on the material combination, see [37,40,41]. For certain material combinations, such as aluminium-steel, a decrease in the lower limit angle with increasing collision velocity was even observed. On the other hand, the upper limit angle shows an increasing curve for all material combinations investigated so far. In the present investigations, a comparatively strong increase in the lower limit angle with increasing velocity was observed, significantly more pronounced than in previous studies. The angle range of the process window at 297 m/s lies between the lower limit angle of 4.7° and the upper limit angle of 5.5°. At a maximum collision speed of 349 m/s, this results in an angle range of 5.8° to 8.4° for the material combination investigated. The physical reasons for the sharp increase in the lower limit angle with increasing speed are not yet fully understood.

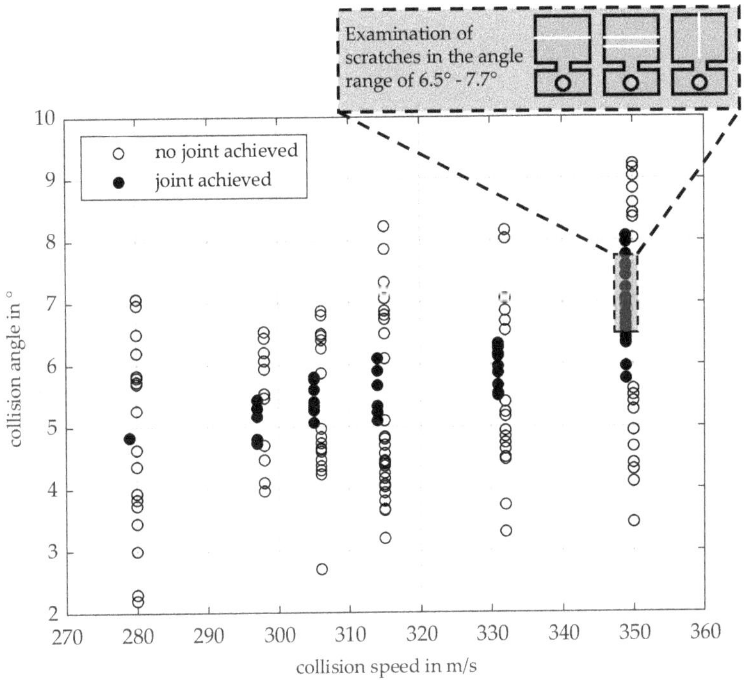

Figure 4. Welding process window of the material combination AA6110-T4 (target, thickness: 2.0 mm) and AA6060-T6 (flyer, thickness: 2.0 mm).

Since the joint quality typically deteriorates as the limit angles are approached, and since the exact setting of the collision angle on the test bench cannot be fully guaranteed due to dynamic effects and tolerance-related deviations (see [37]), an intermediate angle range of 6.5° to 7.7° at 349 m/s was selected for the following investigations of the influence of scratches on joint formation. This choice makes it possible to observe the influence of scratches in an isolated manner under constant and reproducible process conditions.

The area marked in Figure 4 represents the three scratch configurations to be investigated in the described process window area. The complete test plan with all investigated parameter variations is summarised in Table 1.

3.2. Welding Ratio

The results of the ultrasonic examinations of the scratched samples are shown in Figure 5. The left side of the figure shows the qualitatively analysed ultrasonic images of exemplary samples of the respective test configurations. The dotted lines indicate the position of the scratches. The scratch depth increases towards the bottom for all scratch variants.

Areas where the two materials are joined are coloured blue. Red areas indicate areas that are not joined. It is clear that the bonded area decreases significantly with increasing scratch depth in all the configurations analysed. This effect is particularly pronounced for scratches with an orientation of 90° to the welding direction. In the configuration with two parallel scratches and maximum scratch depth, no joint could be achieved within the angle and velocity range considered.

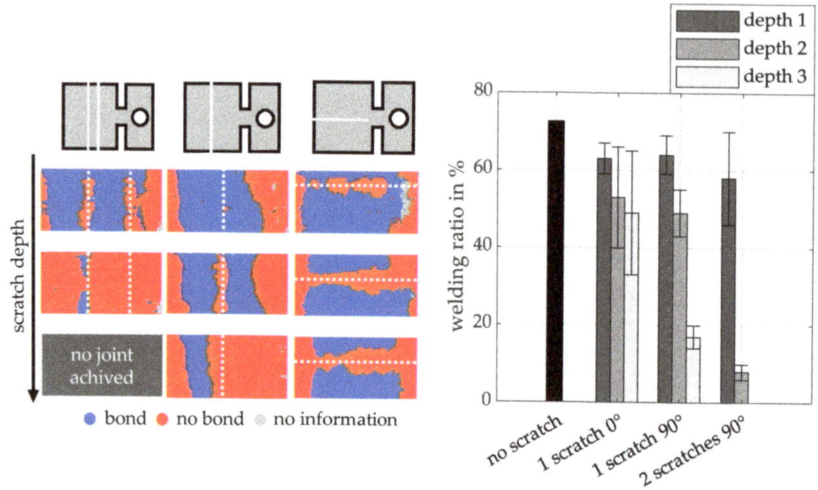

Figure 5. (**left**) Ultrasonic examination of the scratched and welded specimens (welding direction: from right to left). (**right**) Welding ratios of the analysed specimens.

The bar graphs on the right show the quantitative percentage welding ratios derived from the ultrasonic measurements. The welding ratio is defined as the ratio of the welded area to the total overlap area A_W (see Figure 3). For comparison, the welding ratio of a reference sample without scratches and with identical process parameters is also shown.

The results show that scratches with an orientation of 0° to the welding direction have only a moderate effect on the welding ratio compared to the scratch-free reference sample with increasing depth (reduction of 12% at the lowest depth and 24% at the greatest depth). In contrast, the welding ratio decreases significantly more with a 90° scratch orientation, up to 56% at the greatest scratch depth. At the lowest scratch depth, the welding ratio for the 0° and 90° orientations is in the same range. However, increasing the number of scratches has an even more significant effect. Even at the lowest depth, two parallel scratches of identical depth and orientation result in a reduction in welding ratio of 25% compared to the reference sample. At a medium scratch depth, the welding ratio is only 8%. At the greatest scratch depth, as already mentioned, no joint is achieved.

Figure 6 shows a panoramic view of the weld zone of a specimen with two 90° scratches of medium-depth, composed of individual SEM images. The weld direction is analogous to the ultrasonic images shown above, from right to left. The SEM panorama confirms the ultrasonic image. In the area before the first scratch, the whole area is not connected and is characterised by a thick separation layer with inclusions. After the first scratch, the thickness of the separating layer decreases, but the samples are not connected along the length up to the second scratch. In the area of the second scratch, a more porous structure can be observed compared to the base material. The original scratch is completely filled with this microstructure. Small parts or remnants of this porous structure can also be seen in the area of the first scratch. After the second scratch, a connected area can be identified.

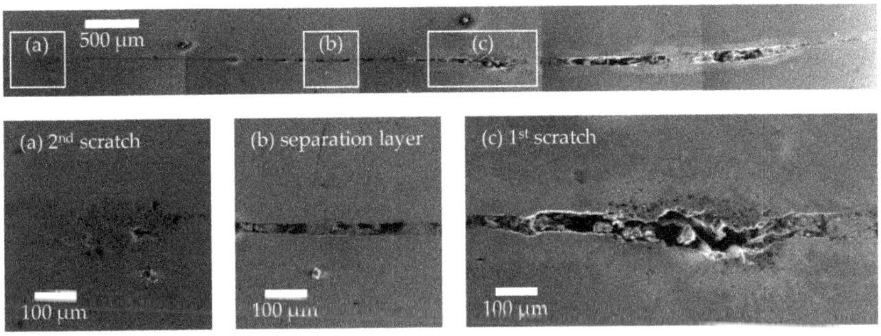

Figure 6. Panoramic view of the weld zone of a specimen with two medium-depth scratches oriented at 90° to the welding direction (on top: target, on the bottom: flyer, welding direction: from right to left, panoramic view stitched together from five separate SEM images). Magnification sections: (**a**) 2nd scratch, (**b**) separation layer with inclusions, (**c**) 1st scratch.

3.3. Tensile Shear Test

Figure 7a shows examples of force-travel curves obtained when determining the tensile shear strength of a scratch-free specimen and of specimens with a scratch oriented at 90° to the welding direction, at three different scratch depths. The curves show the maximum load and travel distance before the joint separates. The ratios of the force values determined do not correspond exactly to the shear strength values of the specimens presented later, since the cutting of the welded specimens is subject to the tolerances described, and therefore, deviations in the maximum force occur to a certain extent. When calculating the tensile shear strength, the determined force is related to the overlapped area A_W. The specimen with the lowest scratch depth already shows a slightly lower endured maximum tensile shear force compared to the scratch-free specimen. As the scratch depth increases, the maximum tensile shear force decreases further, and the welded specimens separate after the testing machine has travelled a shorter distance. A significantly steeper drop in tensile shear force is also observed in specimens with medium and maximum scratch depths. Previous studies [37] have identified two distinct failure mechanisms in tensile shear tests on collision-welded specimens. Those with high tensile shear strength values fail due to shear in the lower-strength base material, i.e., the AA6060-T6 of the flyer in the present material combination (no scratch and scratch depth 1). In contrast, specimens with low tensile shear strength values (scratch depths 2 and 3) fail due to sliding in the weld zone. This sliding after reaching the maximum bearable force occurs almost abruptly, so that the switch-off threshold of the testing machine is reached, as the specimens can no longer transfer any force from this moment on.

The right-hand side of the figure shows examples of completely separated specimens after the tensile shear test. The scratch-free reference specimen (Figure 7b) demonstrates shear failure in the base material over almost its entire length. Only at the beginning and end of the joint does sliding occur in the weld zone (light silver, smooth areas). In the specimen with the lowest scratch depth (Figure 7c), shear failure in the base material also occurs over the largest area. However, sliding in the weld zone is already recognisable in a small partial area near the scratch. At medium scratch depth (Figure 7d), the area in which sliding occurs in the weld zone is predominant. Dark discolouration (white outlined areas) in the area of the scratch indicates that the jet remains there for a longer period of time, or even becomes trapped. At the greatest scratch depth (Figure 7e), dark discolouration (white outlined areas) is observed across almost the entire surface of the weld zone in front of the scratch. Only in the area after the scratch are light, silver-coloured, smooth areas visible, indicating failure due to sliding in the weld zone. Comparing this with the previously shown ultrasonic evaluations reveals that the dark discolourations are present in areas where no joint is formed. The ultrasonic images of the three samples shown in Figure 7c–e, which have a 90° oriented scratch in three scratch depths, are shown in the centre column of Figure 5. Comparing the welded and non-welded areas of the separated specimens with the ultrasonic images reveals a high degree of correspondence.

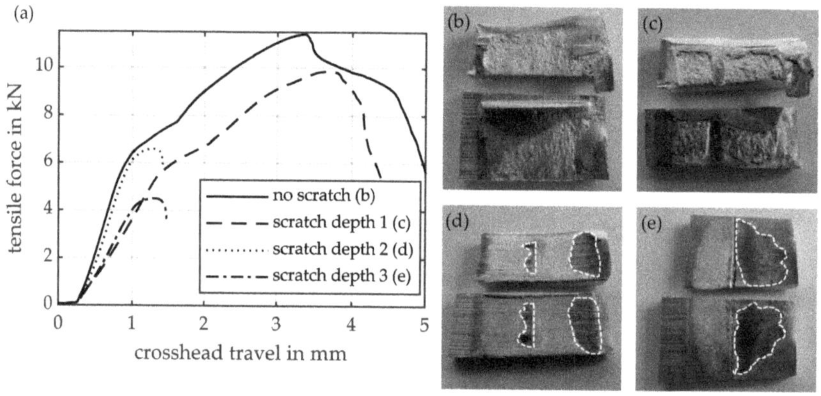

Figure 7. (**a**) Force-travel curves of specimens with and without scratches in the tensile shear test. (**b–e**) Images of the separated specimens after the tensile shear test (on top: flyer, on the bottom: target, welding direction: from right to left).

Figure 8a shows the tensile shear strength of the specimens that were previously analysed using ultrasonic imaging. As can be seen, scratches with a shallow depth (15–30 µm) have no significant influence on the strength of the joint. This applies both to a scratch oriented at 0° to the welding direction and to one or two scratches oriented at 90°. In accordance with the results of the quantitative welding ratios (see Figure 5), the tensile shear strength decreases from the average scratch depth (50–80 µm) in all configurations investigated. The influence of scratch orientation is clearly recognisable in these results. While a moderate decrease in strength can be observed with scratches oriented at 0°, a 90° orientation (i.e., perpendicular to the welding direction) leads to a significantly greater decrease. The lowest strengths occur in samples with two scratches at a 90° angle and a medium scratch depth. With two scratches at the greatest depth, bonding is not achieved.

SEM images of the scratch zones are taken prior to tensile shear testing. Figure 8b shows an example of the structures observed by SEM for a 90° scratch at three different depths. At low scratch depths, only a locally altered microstructure is visible in the immediate vicinity of the scratch. In the welding direction, both before and after the scratch, the joint is fully welded. At a medium scratch depth, the porous structure, characterised by cavities in the area of the scratch, indicates vortex formation. This suggests turbulence in the escaping jet and entrained material, as well as complex material rearrangements. With a diameter of over 300 µm, the swirled area is significantly greater than the original scratch depth. This means that material melts directly around the scratch during the collision. This structure is also observed in the area directly in front of and behind the scratch, but to a lesser extent. The SEM image of the deepest scratch shows residues of possible vortex formation during the collision. However, compared to the medium scratch depth, there is almost no material in the scratch, and the localised melting around the scratch is smaller in relation to the original scratch depth. Additionally, the materials in the areas before and after the scratch are not connected. The areas on the left and right edges of the scratch in the image suggest that the samples may have been temporarily connected before being separated by a crack. Regarding the SEM images and observations, it should be noted that they are exemplary, as they only permit localised observation within the weld zone. It is not possible to observe the entire length of the scratch.

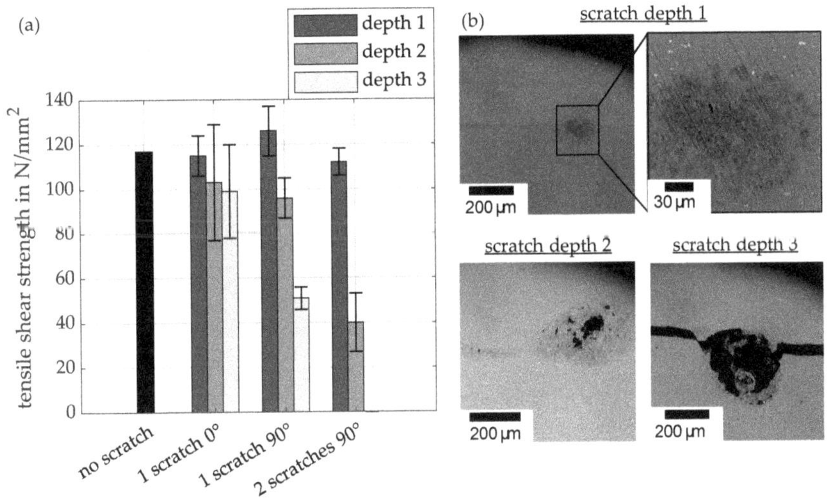

Figure 8. (**a**) Tensile shear strength of the analysed specimens. (**b**) SEM images of the scratch zones at the three different scratch depths.

3.4. High-Speed Images

Figure 9 shows an example of the high-speed images taken during the collision. The images compare a scratch-free reference specimen with a specimen prepared with two medium-depth scratches oriented at 90° to the welding direction. The resulting jet is clearly visible at the beginning of the collision in both sequences of images. The scratches in the prepared specimen are clearly visible in the first image. As the collision progresses, the untreated specimen exhibits the characteristic formation of the jet. This spreads continuously throughout the collision process as the collision gap closes, leaving the collision gap along the collision direction at the end. In contrast, lateral jet clouds form in the specimen with the two scratches. These form in the area of the scratches, or are deflected there from the jet's original direction of travel. Although a jet is also present in the collision direction, its character and intensity are significantly reduced compared to the reference specimen. The middle image sequence, in particular, shows that the jet is more intense and focused in the non-scratched specimen.

An additional specimen is produced to investigate this phenomenon in more detail. In this specimen, the bore of the target for mounting on the rotor is offset by 1.0 mm laterally in order to deliberately create a lateral offset between the colliding specimens during the collision. This specimen is welded using the same process parameters as before (v_{imp} = 349 m/s, β = 7.2°). A dark, localised deposit similar to the discolourations shown in Figure 7 can be seen in the area of the scratch. This confirms the assumption, based on the high-speed images, that the jet emerges from the side of the sample in the area of the scratches.

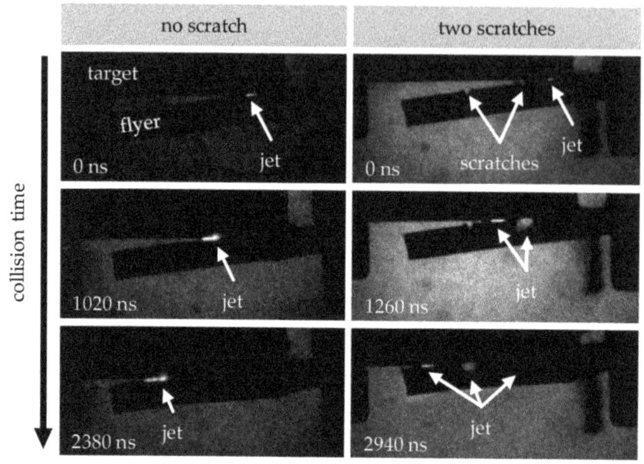

Figure 9. High-speed images taken during the collision of a specimen at the model test rig (welding direction: from right to left, v_{imp} = 349 m/s). (**left**) Specimen with no scratches. (**right**) Specimen with two scratches of scratch depth 2 oriented at 90°.

4. Discussion

The results presented demonstrate how different scratch configurations influence the quality of joint formation during collision welding of two aluminium specimens made from AA6060-T6 and AA6110A-T6 alloys. The effects of scratches of different depths and orientations relative to the welding direction on the welding result are investigated. These scratches lead to varying welding ratios and interlayers, directly affecting the resulting tensile shear strength.

The main reason for the observed differences in strength values is the direct correlation between the welding ratio and the calculated shear strength. As the latter is related to the complete overlap area A_w (see Figure 3), reducing the welded proportion (with constant strength per surface element) will result in a drop in the calculated tensile shear strength. This relationship is clearly illustrated in Figures 5 and 8. It can be seen that shallow scratches with a depth of 15–30 µm only slightly reduce the strength values compared to scratch-free reference samples. The lowest strengths are observed for deep scratches at an angle of 90°.

Previous studies have shown that a jet consisting of removed oxide layers, impurities, and ionised material, together with high surface pressure, is essential for forming a high-quality joint. The jet both cleans and activates the surfaces. [34] Based on the results obtained in this study, it can be assumed that scratches above a certain depth negatively affect the formation and propagation of the jet, thereby reducing the welding ratio. This assumption is supported by the high-speed images in Figure 9, which show disturbed jet formation, and the cross-sections in Figures 6 and 8. Two characteristic scenarios can be distinguished.

1. Jet inclusion in the scratch: As the collision progresses, the jet hits the scratch and becomes part of it. Depending on the size of the scratch, it is either completely or partially filled. The resulting heat, which has also been observed in collision welding tests at low collision angles [37], can cause the material to melt locally. Combined with the removed material, this creates a pronounced intermediate layer. Similar findings were reported in EMPW welding processes involving ground surfaces [42]. In particular, structures running perpendicular to the welding direction lead to increased interlayer formation, which can be attributed to jet inclusions.

2. Jet interruption due to scratch geometry: In some cases, the jet is deflected or interrupted so strongly by the scratches (the indentation and the elevations at the edges of the scratches, see scratch profile in Figure 1) that there is insufficient energy left for surface cleaning and activation. SEM images in Figure 8 show vortex structures indicating such jet disturbance. Notably, this prevents joint formation in large areas before the scratch, whereas a joint is formed after the scratch. This suggests strong turbulence in the closing collision gap, hindering the jet and connection formation in the area in front of the scratch. The turbulent atmosphere escapes from both sides of the collision gap via the scratch, creating atmospheric conditions again after the scratch that enable joint formation.

Compared to the 90° scratches described above, those in a 0° orientation have less influence on jet and joint formation. With these scratches, the sample is fully joined in the areas outside the scratch area, regardless of the depth of the scratch. However, in the scratch area, there is almost no joint formation for the three scratch depths. This may be due to a lack of material being pressed into the 0° oriented scratch during the collision. The local plastic deformation and the local material flow are not sufficient to form a joint. On the other hand, for a 90° oriented scratch, the possibility of material being pressed into the scratch is greater due to the rolling motion of the flyer on the target surface during the collision. Nevertheless, some joints do occur in certain areas at low scratch depths, due to sufficient plastic deformation and local material flow within the scratch area. This clearly demonstrates the sensitivity of the process to external influences. At higher scratch depths, there are no joined areas within the scratch path. From a certain depth onwards (compare ultrasonic examination in Figure 5), it is no longer possible to smooth out the scratches through the collision process. This results in the complete elimination of joint formation in these areas.

5. Conclusions

This study determines and analyses the process window of the material combination AA6060-T6 (extruded flat material, thickness: 2.0 mm) and AA6110A-T6 (extruded flat material, thickness: 2.0 mm) using a model test rig for collision welding. Based on this, the influence of different types of scratches on the joint quality of collision-welded samples is analysed systematically. The scratches vary in terms of depth, orientation, and number. The results clearly demonstrate that increasing both the scratch depth and number leads to a significant reduction in the tensile shear strength that the joint can endure. Scratches oriented perpendicular to the collision direction have a particularly critical effect as they lead to turbulence in the jet from a critical scratch depth of 50–80 μm, which is accompanied by a significantly greater reduction in tensile shear strength. This is mainly due to a reduction in the welding ratio. The influence of partially formed intermediate layers on tensile shear strength could not be quantified in this study, so this is an important area for future research.

In addition to further investigating the varying increases in limit angles with different material combinations within their respective welding process windows, future research should analyse the influence of scratches on the resulting intermediate layers more in-depth. As part of an extended parameter study, the influence of other scratch configurations should also be considered to develop a comprehensive catalogue of influencing parameters for industrial applications. This will enable critical threshold values to be identified with regard to scratch depth and number for specific applications. This will enable targeted countermeasures to be derived in the area of material handling, as well as improved control of surface-related quality fluctuations. These measures can be adapted as required, ultimately reducing effort and costs in the industrial production process.

Author Contributions: Conceptualization, S.O.K., J.B., F.S. and P.G.; methodology, S.O.K., J.B., F.S. and P.G.; software, S.O.K., J.B. and F.S.; validation, S.O.K., J.B. and F.S.; formal analysis, S.O.K., J.B. and F.S.; investigation, S.O.K., J.B. and F.S.; resources, P.G.; data curation, S.O.K., J.B. and F.S.; writing—original draft preparation, S.O.K. and J.B.; writing—review and editing, S.O.K., J.B. and P.G.; visualization, S.O.K. and J.B.; supervision, P.G.; project administration, S.O.K.; funding acquisition, P.G. All authors have read and agreed to the published version of the manuscript.

Funding: This research is a result of the project Design2Collide—Kollisionsgeschweißte Hybrid-Leichtbaustrukturen. The project is funded by the Federal Ministry for Economic Affairs and Energy (BMWE, 03LB5008H), based on a decision taken by the German Bundestag. The authors are solely responsible for the content in this publication.

Institutional Review Board Statement: Not applicable.

Informed Consent Statement: Not applicable.

Data Availability Statement: The original data presented in the study are openly available in [43].

Conflicts of Interest: The authors declare no conflicts of interest. The funders had no role in the design of the study; in the collection, analyses, or interpretation of data; in the writing of the manuscript, or in the decision to publish the results.

References

1. Miraftabzadeh, S.M.; Longo, M.; Foiadelli, F. Mobility and Future Trends. In *Emerging Battery Technologies to Boost the Clean Energy Transition*, 1st ed.; Passerini, S., Barelli, L., Baumann, M., Peters, J., Weil, M., Eds.; Springer: Cham, Switzerland, 2024; Volume 1, pp. 3–11. [CrossRef]
2. Crippa, M.; Guizzardi, D.; Pagani, F.; Banja, M.; Muntean, M.; Schaaf, E.; Monforti-Ferrario, F.; Becker, W.; Quadrelli, R.; Risquez Martin, A.; et al. *GHG Emissions of all World Countries 2024*; Publications Office of the European Union: Luxembourg, 2024. [CrossRef]
3. EDGAR (Emissions Database for Global Atmospheric Research). IEA-EDGAR Fossil CO_2 Emissions. 2024. Available online: https://edgar.jrc.ec.europa.eu/report_2024#data_download (accessed on 7 April 2025).
4. IEA (International Energy Agency). Global Total Final Consumption by Fuel in the Net Zero Scenario, 2010–2050. 2025. Available online: www.iea.org/data-and-statistics/charts/global-total-final-consumption-by-fuel-in-the-net-zero-scenario-2010-2050 (accessed on 7 April 2025).
5. Timmers, V.R.J.H.; Achten, P.A.J. Non-exhaust PM emissions from electric vehicles. *Atmos. Environ.* **2016**, *134*, 10–17. [CrossRef]
6. Lopez, B.; Wang, X.; Chen, L.W.A.; Ma, T.; Mendez-Jimenez, D.; Cobb, L.C.; Frederickson, C.; Fang, T.; Hwang, B.; Shiraiwa, M.; et al. Metal contents and size distributions of brake and tire wear particles dispersed in the near-road environment. *Sci. Total Environ.* **2023**, *883*, 163561. [CrossRef]
7. European Environment Agency (EEA). Air Quality in Europe 2022. Available online: https://www.eea.europa.eu/publications/air-quality-in-europe-2022 (accessed on 7 April 2025).
8. Köllner, C. Unterschätzte Umweltgefahr Reifenabrieb. Available online: https://www.springerprofessional.de/fahrwerk/schadstoffe/unterschaetzte-umweltgefahr-reifenabrieb-/15490524 (accessed on 7 April 2025).
9. Organisation for Economic Co-Operation and Development (OECD). Non-Exhaust Particulate Emissions from Road Transport. Available online: https://doi.org/10.1787/4a4dc6ca-en (accessed on 7 April 2025).
10. European Commission. Questions and Answers: Commission Proposal on the New Euro 7 Standards. Available online: https://ec.europa.eu/commission/presscorner/detail/en/QANDA_22_6496 (accessed on 7 April 2025).
11. Allgemeiner Deutscher Automobil-Club e.V. (ADAC). Euro-7-Abgasnorm: Anforderungen, Grenzwerte, Einführungsfristen. Available online: https://www.adac.de/rund-ums-fahrzeug/autokatalog/abgasnormen/euro-7 (accessed on 7 April 2025).
12. Visnic, B. Europe's Dust Buster. Available online: https://www.sae.org/news/2023/05/brakes-sustainability-dust-regulations (accessed on 7 April 2025).
13. Kole, P.J.; Löhr, A.J.; Van Belleghem, F.G.; Ragas, A.M. Wear and Tear of Tyres: A Stealthy Source of Microplastics in the Environment. *Int. J. Environ. Res. Public Health* **2017**, *14*, 1265. [CrossRef] [PubMed]
14. The International Transport Forum (OECD). Lightening Up: How Less Heavy Vehicles Can Help Cut CO2 Emissions. Available online: https://www.itf-oecd.org/less-heavy-vehicles-cut-co2-emissions (accessed on 7 April 2025).
15. Bandivadekar, A.; Bodek, K.; Cheah, L.; Evans, C.; Groode, T.; Heywood, J.; Kasseris, E.; Kromer, M.; Weiss, M. *On the Road in 2035: Reducing Transportation's Petroleum Consumption and GHG Emissions*; MIT Laboratory for Energy and the Environment: Cambridge, MA, USA, 2008.

16. Köllner, C. So Lassen Sich Brems- und Reifenabrieb Reduzieren. Available online: https://www.springerprofessional.de/fahrwerk/partikel---feinstaub/so-lassen-sich-brems--und-reifenabrieb-reduzieren/18816284 (accessed on 7 April 2025).
17. Beddows, D.C.S.; Harrison, R.M. PM10 and PM2.5 emission factors for non-exhaust particles from road vehicles: Dependence upon vehicle mass and implications for battery electric vehicles. *Atmos. Environ.* **2021**, *244*, 117886. [CrossRef]
18. Tisza, M.; Lukács, Z. High strength aluminum alloys in car manufacturing. *IOP Conf. Ser. Mater. Sci. Eng.* **2018**, *418*, 012033. [CrossRef]
19. Kang, M.; Kim, C. A Review of Joining Processes for High Strength 7xxx Series Aluminum Alloys. *J. Weld. Join.* **2017**, *35*, 79–88. [CrossRef]
20. Olabode, M.; Kah, P.; Martikainen, J. Aluminium alloys welding processes: Challenges, joint types and process selection. *Proc. Inst. Mech. Eng. Part B J. Eng. Manufact.* **2013**, *227*, 1129–1137. [CrossRef]
21. Tao, Y.; Luo, S.; Zhang, J.; Wang, X.; Chen, X.; Hu, Z.; Nagaumi, H. Effect of aging state on microstructure and properties of heat affected zone in Al–Mg–Si–Cu alloy welded joints. *J. Mater. Res. Technol.* **2024**, *33*, 3045–3057. [CrossRef]
22. MachineMFG. Aluminum Welding: Techniques, Challenges, and Applications. Available online: https://shop.machinemfg.com/aluminum-welding-techniques-challenges-and-applications (accessed on 6 May 2025).
23. Agudo, L.; Eyidi, D.; Schmaranzer, C.H.; Arenholz, E.; Jank, N.; Bruckner, J.; Pyzalla, A.R. Intermetallic Fe_xAl_y-phases in a steel/Al-alloy fusion weld. *J. Mater. Sci.* **2007**, *42*, 4205–4214. [CrossRef]
24. Khedr, M.; Hamada, A.; Järvenpää, A.; Elkatatny, S.; Abd-Elaziem, W. Review on the Solid-State Welding of Steels: Diffusion Bonding and Friction Stir Welding Processes. *Metals* **2023**, *13*, 54. [CrossRef]
25. Kapil, A.; Sharma, A. Magnetic pulse welding: An efficient and environmentally friendly multi-material joining technique. *J. Clean. Prod.* **2015**, *100*, 35–58. [CrossRef]
26. Niessen, B. Untersuchungen zur Robustheit von Kollisionsschweißprozessen bei Minimalem Energieeintrag. Ph.D. Thesis, TU Darmstadt, Darmstadt, Germany, 2022. [CrossRef]
27. Atabaki, M.M.; Nikodinovski, M.; Chenier, P.; Ma, J.; Harooni, M.; Kovacevic, R. Welding of Aluminum Alloys to Steels: An Overview. *J. Manuf. Sci. Prod.* **2014**, *14*, 59–78. [CrossRef]
28. Golovashchenko, S. Electromagnetic Forming and Joining for Automotive Applications. In Proceedings of the 2nd international Conference on High Speed Forming, Dortmund, Germany, 20–21 March 2006; Kleiner, M., Ed.; Institute of Forming Technology and Lightweight Construction, Universität Dortmund: Dortmund, Germany, 2006; Volume 1, pp. 201–206. [CrossRef]
29. Weddeling, C.; Woodward, S.; Nellesen, J.; Psyk, V.; Marré, M.; Brosius, A.; Tekkaya, A.E.; Daehn, G.S.; Tillmann, W. Development of design principles for form-fit joints in lightweight frame structures. In Proceedings of the 4th International Conference on High Speed Forming, Columbus, OH, USA, 9–10 March 2010; Babusci, K., Daehn, G., Marré, M., Tekkaya, A.E., Weddeling, C., Zhang, Y., Eds.; Manufacturing Institute, The Ohio State University: Columbus, OH, USA, 2010; Volume 1, pp. 137–148. [CrossRef]
30. Wang, X.; Li, C.; Zhou, Y.; Xu, C.; Dai, M. Mechanism of the discharge behavior in electromagnetic pulse welding: Combination of electron emission and electric field. *Mater. Today Commun.* **2023**, *36*, 106726. [CrossRef]
31. Stankevic, V.; Lueg-Althoff, J.; Hahn, M.; Tekkaya, A.E.; Zurauskiene, N.; Dilys, J.; Klimantavicius, J.; Kersulis, S.; Simkevicius, C.; Balevicius, S. Magnetic Field Measurements during Magnetic Pulse Welding Using CMR-B-Scalar Sensors. *Sensors* **2020**, *20*, 5925. [CrossRef]
32. Marschner, O.; Pabst, C.; Schäfer, R.; Pasquale, P. Suitable Design for Electromagnetic Pulse Processes. In Proceedings of the 9th International Conference on High Speed Forming, Online, 13–15 October 2021; Institute of Forming Technology an Lightweight Construction, Universität Dortmund: Dortmund, Germany, 2021; Volume 1, pp. 1–9. [CrossRef]
33. Zerbst, U.; Madia, M.; Klinger, C.; Bettge, D.; Murakami, Y. Defects as a root cause of fatigue failure of metallic components. III: Cavities, dents, corrosion pits, scratches. *Eng. Fail. Anal.* **2019**, *97*, 759–776. [CrossRef]
34. Bellmann, J.; Lueg-Althoff, J.; Niessen, B.; Böhme, M.; Schumacher, E.; Beyer, E.; Leyens, C.; Tekkaya, A.E.; Groche, P.; Wagner, M.F.-X.; et al. Particle Ejection by Jetting and Related Effects in Impact Welding Processes. *Metals* **2020**, *10*, 1108. [CrossRef]
35. Deutsches Institut für Normung e. V. *DIN EN 573-3 Aluminium and Aluminium Alloys—Chemical Composition and form of Wrought Products—Part 3: Chemical Composition and form of Products*; German Version EN 573-3:2019+A2:2023; Beuth Verlag GmbH: Berlin, Germany, 2024.
36. Deutsches Institut für Normung e. V. *DIN 50125 Testing of Metallic Materials—Tensile Test Pieces*; Beuth Verlag GmbH: Berlin, Germany, 2022.
37. Kraus, S.O.; Bruder, J.; Groche, P. The Influence of Weld Interface Characteristics on the Bond Strength of Collision Welded Aluminium–Steel Joints. *Materials* **2024**, *17*, 3863. [CrossRef]
38. Tsakiris, V.; Kappel, W.; Alecu, G. Solid State diffusion welding of Cu-Fe/Al/Ag and Al-Ni dissimilar metals. *J. Optoel. Adv. Mat.* **2011**, *13*, 1176–1180. [CrossRef]
39. Böhme, M.; Sharafiev, S.; Schumacher, E.; Böhm, S.; Wagner, M.F.X. On the microstructure and the origin of intermetallic phase seams in magnetic pulse welding of aluminium and steel. *Materialwiss. Werkstofftech.* **2019**, *50*, 958–964. [CrossRef]

40. Niessen, B.; Schumacher, E.; Lueg-Althoff, J.; Bellmann, J.; Böhme, M.; Böhm, S.; Tekkaya, A.E.; Beyer, E.; Leyens, C.; Wagner, M.F.-X.; et al. Interface Formation during Collision Welding of Aluminum. *Metals* **2020**, *10*, 1202. [CrossRef]
41. Groche, P.; Niessen, B. The Energy Balance in Aluminum–Copper High-Speed Collision Welding. *J. Manuf. Mater. Process.* **2021**, *5*, 62. [CrossRef]
42. Cui, J.; Sun, T.; Geng, H.; Yuan, W.; Li, G.; Zhang, X. Effect of surface treatment on the mechanical properties and microstructures of Al-Fe single-lap joint by magnetic pulse welding. *Int. J. Adv. Manuf. Technol.* **2018**, *98*, 1081–1092. [CrossRef]
43. Kraus, S.O.; Bruder, J.; Schuller, F.; Groche, P. *High-Speed Images of Process Window Determination for AA6110-T6—AA6060-T6 Material Combination in Collision Welding*; TUdata, TU Darmstadt: Darmstadt, Germany, 2025. [CrossRef]

materials

MDPI

Article

Modeling and Experimental Investigation of the Impact of the Hemispherical Tool on Heat Generation and Tensile Properties of Dissimilar Friction Stir Welded AA5083 and AA7075 Al Alloys

Ahmed R. S. Essa [1,2], Ramy I. A. Eldersy [1], Mohamed M. Z. Ahmed [3,4], Ali Abd El-Aty [3,5,*], Ali Alamry [3], Bandar Alzahrani [3], Ahmed E. El-Nikhaily [1] and Mohamed I. A. Habba [1]

1 Mechanical Department, Faculty of Technology and Education, Suez University, Suez 43211, Egypt; ahmed.eessa@suezuniv.edu.eg (A.R.S.E.); iramy3791@gmail.com (R.I.A.E.); ahmed.eassa@ind.suezuni.edu.eg (A.E.E.-N.); mohamed.atia@suezuniv.edu.eg (M.I.A.H.)
2 Faculty of Engineering, King Salman International University, El Tor 45615, Egypt
3 Department of Mechanical Engineering, College of Engineering at Al Kharj, Prince Sattam Bin Abdulaziz University, Riyadh 11942, Saudi Arabia; moh.ahmed@psau.edu.sa (M.M.Z.A.); a.alamry@psau.edu.sa (A.A.); ba.alzahrani@psau.edu.sa (B.A.)
4 Department of Metallurgical and Materials Engineering, Faculty of Petroleum and Mining Engineering, Suez University, Suez 43211, Egypt
5 Mechanical Engineering Department, Faculty of Engineering, Helwan University, Cairo 11795, Egypt
* Correspondence: a.hassibelnaby@psau.edu.sa

Abstract: This study investigated the effect of a hemispherical friction stir welding (FSW) tool on the heat generation and mechanical properties of dissimilar butt welded AA5083 and AA7075 alloys. FSW was performed on the dissimilar aluminum alloys AA5083-H111 and AA7075-T6 using welding speeds of 25, 50, and 75 mm/min. The tool rotation rate was kept constant at 500 rpm. An analytical model was developed to calculate heat generation and temperature distribution during the FSW process utilizing a hemispherical tool. The experimental results were compared to the calculated data. The latter confirms the accuracy of the analytical model, demonstrating a high degree of agreement. Sound FSW dissimilar joints were achieved at welding speeds of 50 and 25 mm/min. Meanwhile, joints created at a welding speed of 75 mm/min exhibited a tunnel-like defect, which can be attributed to the minimal heat generated at this particular welding speed. At a lower welding speed of 25 mm/min, a higher tensile strength of the dissimilar FSWed joints AA5083 and AA7075 was achieved with a joint efficiency of over 97%.

Keywords: friction stir welding; hemispherical tool; peak temperature; dissimilar aluminum alloys; mechanical properties

1. Introduction

Since its invention in 1991, friction stir welding (FSW) has gained popularity for joining different aluminum alloys (AA), particularly difficult-to-weld alloys such as the 2xxx and 7xxx series [1]. FSW parameters, tool design, and tool materials are continuously updated. Process operation, including the development of the bobbin tool [2–6], stationary shoulder [7–9], reverse dual rotation [10], and others, has also progressed. Recent advancements in solid-state welding have made FSW one of the best methods for joining various alloys, particularly dissimilar alloys. Several process parameters control the FSW of dissimilar alloys. The impact of various factors on the characteristics of friction stir-welded joints between dissimilar aluminum alloys has been the subject of numerous studies [8–12]. Tool geometry and design have attracted the attention of many researchers because of the vital role that the tool plays in the quality of friction stir welded joints [13–19]. According to Raj Kumar et al. [20], FSW of dissimilar aluminum alloys AA5052 and AA6061 is affected by welding settings and tool design. They reported that the alloys may be joined effectively and with good bonding using cylindrical threaded pins. Compared with other tools, the stepped pin tool provided the maximum tensile strength for the welded joints, and generated a sound joint at welding speeds of 1000 rpm and 40 mm/min. A study [1] on the effect of pin eccentricity on the characteristics of friction stir welded AA7075 found that a clear impact on the tensile characteristics of the welded joints; a pin eccentricity of 0.2 mm is the ideal value for maximizing strength and elongation.

In the FSW process, the heat generated is assumed to be a combination of two sources: (i) the friction between the tool and the workpiece and (ii) the plastic shear deformation of the weld nugget in the vicinity of the pin [21–25]. The tool contacts the materials and generates the needed heat through three regions: the shoulder, the pin side, and the pin tip, so the friction between the pin side and the joint material is an important source of heat. Moreover, increasing the pin side area will greatly affect the share of every part of the FSW tool in generating heat by friction [26–30]. Sadoun et al. [31] studied the effect of pin side area on heat generation and macrostructure developed during friction stir welding 5 mm thick AA7075 joints. They reported that the temperature distribution is influenced by the ratio of the pin side area, as it directly impacts the quantity of heat created by the pin side. Therefore, a tool with a higher ratio of pin side area is favored from a microstructure perspective. Employing a tool equipped with a semi-spherical pin and a high pin side area ratio (29.83%) results in a joint characterized by a smaller grain size and greater tensile strength. The joint soundness and quality of dissimilar welding of AA7075-T6 and AA5083-H111 depend mainly on the placement of AA7075 (higher flow stress alloy), which needs a higher heat input than AA5083 [32]. The joint efficiency is also higher at higher heat inputs (low travel speeds). The present work aims to increase the contact area of the pin side by using a modified tool design (hemispherical tool) and to investigate its effect on the properties of dissimilar AA7075/AA5083 friction stir welded joints. An analytical model for heat generation will also be modified according to the current tool design.

2. Materials and Methods

2.1. Material Description

A friction stir welding process was used to create dissimilar joints between the aluminum alloys AA5083-H111 and AA7075-T6, utilizing FSW equipment for testing and development. Tables 1 and 2 list the chemical composition and mechanical characteristics of each alloy. The composition of the AA5083-H111 and AA7075-T6 plates was determined using Foundry-Master Pro (Oxford Instruments, Abingdon, UK). The weld samples were created using two 5 mm thick, 100 mm wide, and 150 mm long plates.

Table 1. Chemical composition of the present aluminum alloys.

Type	Element (wt.%)								
	Si	Fe	Cu	Mn	Mg	Cr	Zn	Ti	Al
5083	0.04	0.15	0.02	0.56	4.75	0.05	0.04	0.05	rest
7075	0.07	0.21	1.94	0.05	2.66	0.21	5.94	0.01	rest

Table 2. Mechanical properties of the base aluminum alloys.

Alloy	Tensile Strength (MPa)	Proof Stress 0.2% (MPa)	Elongation (%)
AA5083-H111	307	156	19
AA7075-T6	571	495	11.5

2.2. FSW Procedures

The welding tools were made from a 30 mm diameter W302 cold-worked tool steel rod (Bühler AG, Branch office Cairo, Cairo, Egypt), that had been heat-treated to HRC62, with the following chemical composition: 0.39% C, 0.1% Si, 0.40% Mn, 5.2% Cr, 0.95% V,

1.4% Mo, and 90.6 wt% Fe. As illustrated in Figure 1, a hemispherical tool pin was made with an 8 mm diameter, 4.6 mm long, and a smooth 19 mm diameter shoulder. The pin also has a 2° concavity.

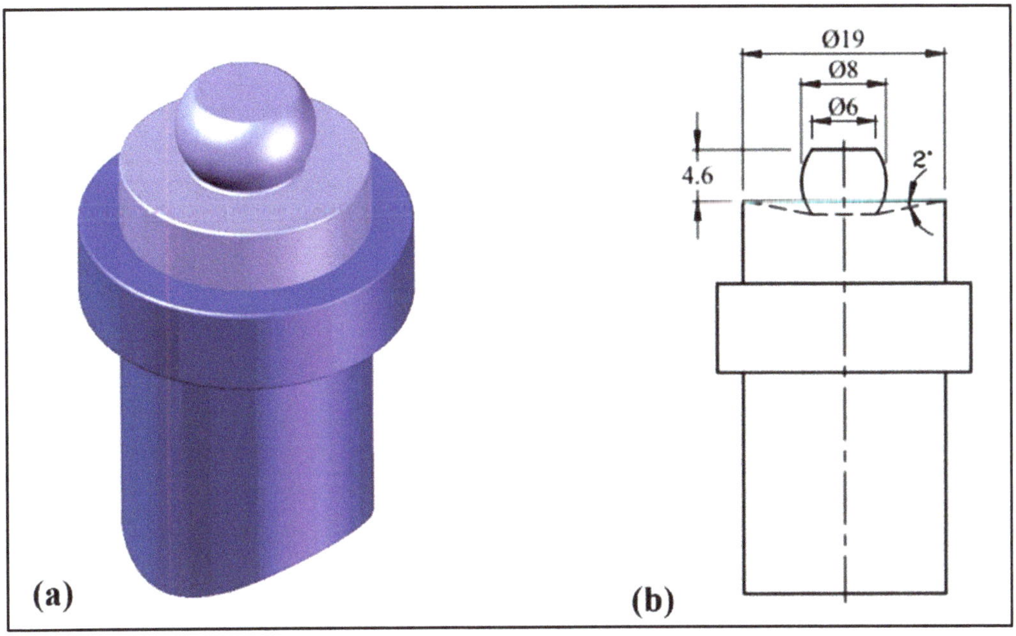

Figure 1. FSW tool (**a**) shape and (**b**) dimensions.

A constant tool plunge depth (shoulder penetration) of 0.2 mm for all experiments was used, and a tilt angle of 3° applied. The peak temperatures were measured on the top surface of the advancing side of the weld joints using a laser thermometer (Quicktemp 8630-T3 device, Testo Company, Berlin, Germany) that has a measurement error range from 1° to 12° based on the object emissivity. The plunge forces were recorded by the FSW machine for research and development. Table 3 lists the applied FSW to produce AA5083-H111 and AA7075-T6 dissimilar joints.

Table 3. The FSW process parameters.

Welding Parameters	Welding Speed (mm/min)	Rotation Speed (rpm)
1	25	500
2	50	500
3	75	500

The metallographic samples were cut perpendicular to the welding direction. The samples were ground, polished, and etched by Keller's chemical agent (2 mL HF, 3 mL HCL, 5 mL nitric acid, and 190 mL water) for 25 s to reveal the microstructure of the weld zones. The grain size developed microstructures were measured using ImageJ software (Version 1.54d, Wayne Rasband and contributors, National Institutes of Health, Bethesda, MD, USA). The Vickers hardness measurements were done using a Vickers Hardness Tester type HWDV-75 (TTS Unlimited, Osaka, Japan). The measurements were made at the center of the joint on the cross-sections perpendicular to the direction of the FSW path. Tensile test specimens were obtained in a direction perpendicular to the welding direction and were prepared in accordance with the ASTM E8/E8M-16a standard. At room temperature and 0.02 mm/s crosshead speed, a universal tensile testing machine (Instron 4208, 30-ton capacity, Norwood, MA, USA) performed tensile tests.

3. Heat Generation Estimation Model

An analytical model for heat generation during friction stir welding was created using a hemispherical pin profile. The suggested model is a revision of earlier analytic models [33,34]. Figure 2a shows the tool design with a hemispherical pin profile and illustrates the three areas where heat is anticipated to be produced by friction. Q_1 is the heat produced by the concave shoulder, Q_2 by the hemispherical side of the pin, and Q_3 by the pin tip, resulting in the overall heat generation, which is given by Equation (1) as:

$$Q_{total} = Q_1 + Q_2 + Q_3 \quad (1)$$

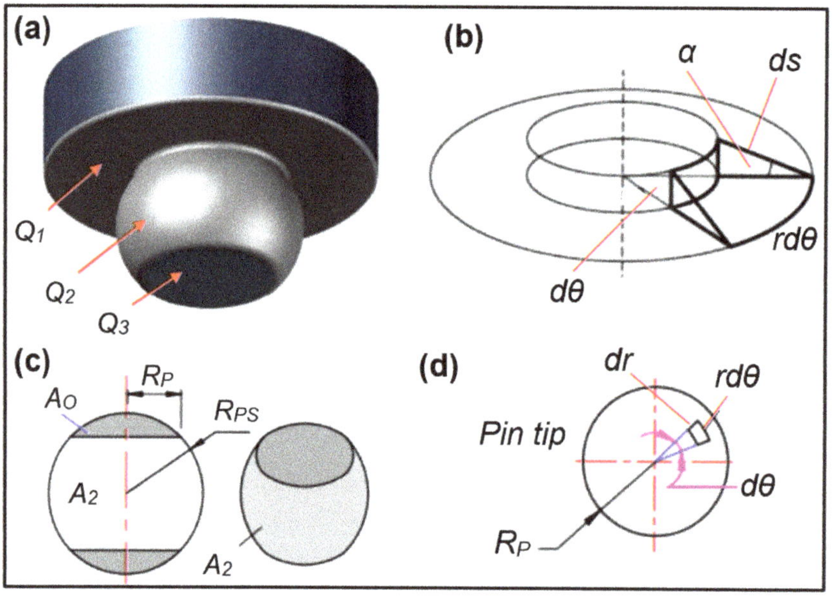

Figure 2. Schematic drawing of surface orientations and infinitesimal segment areas: (**a**) the different heat generation regions, (**b**) concave shoulder, (**c**) pin side, and (**d**) pin tip.

For the current analytical modeling, some assumptions were taken into account. First, the analytical estimation based on the supposition that the contact shear stress is uniform was considered. Second, the sliding state of the shearing occurs at the contact interface. Third, other heat-generating mechanisms, such as deformation, were not considered. Figure 2 depicts the hemispherical tool and workpiece contact surface determined by position and orientation in relation to the rotational axis.

The current analytical modeling, a modified version of the analytical model provided by Essa et al. [35] and Schmidt et al. [36], utilized a straightforward tool design with a concave shoulder surface, a hemispherical pin, and a flat pin surface. The radius of the sphere shape Rps and the radius of the pin tip Rp define the hemispherical pin surface and the concave shoulder surface, respectively. Despite being different, the formulas for each surface area orientation are based on the fundamental equation for heat generation [37]:

$$dQ = \omega dM = \omega \cdot r \cdot dF = \omega \cdot r \cdot \tau_{contact} dA \quad (2)$$

3.1. Heat Generated from the Shoulder Surface

To calculate the heat produced by the concave shoulder surface spinning around the tool axis, an infinitesimal segment on that surface was taken into account. This infinitesimal segment region is subject to uniform contact shear stress, as seen in Figure 2b. This section only makes a tiny force and torque contribution. The heat produced by this section is:

$$dQ_1 = \omega \cdot r \cdot \tau_{contact} r d\theta ds \tag{3}$$

where r is the distance between the considered area and the center of rotation, ω is the angular velocity, and "$d\theta$, ds" are the segment dimensions. Integration of Equation (3) over the concave shoulder area from Rp to Rs gives the shoulder heat generation, Q_1, as follows:

$$dQ_1 = \omega \cdot r^2 \cdot \tau_{contact} \cdot d\theta \cdot \frac{dr}{cos\alpha} \tag{4}$$

$$Q_1 = \int_0^{2\pi} \int_{R_p}^{R_s} \omega \cdot r^2 \cdot \tau_{contact} \cdot d\theta \cdot \frac{dr}{cos\alpha} \tag{5}$$

$$Q_1 = 2\pi \cdot \omega \cdot \tau_{contact} \cdot \frac{\left(R_s{}^3 - R_p{}^3\right)}{3cos\alpha} \tag{6}$$

3.2. Heat Generated from the Hemispherical Surface

The pin consists of a hemispherical surface with a radius of sphere Rps and radius of pin tip Rp and pin height Hp. The heat generated from the pin side is given by the following equations over the pin side area.

$$dQ_2 = \omega \cdot r \cdot \tau_{contact} \cdot dA_2 \tag{7}$$

$$Q_2 = \int \omega \cdot r \cdot \tau_{contact} \cdot dA_2 = \omega \cdot R_{PS} \cdot \tau_{contact} \cdot A_2 \tag{8}$$

$$A_2 = A_{spher} - 2A_O \tag{9}$$

$$A_2 = 4\pi R_{PS}^2 - \left[2\int_0^{2\pi} \int_0^{r_p} \sqrt{1 + \frac{r^2cos^2\theta + r^2sin^2\theta}{r_{PS}^2 - r^2cos^2\theta - r^2sin^2\theta}} rdrd\theta\right] \tag{10}$$

$$A_2 = 4\pi R_{PS}^2 - \left[4\pi R_{PS}^2 - 4\pi R_{PS}\sqrt{R_{PS}^2 + R_P^2}\right] \tag{11}$$

$$A_2 = 4\pi R_{PS}\sqrt{R_{PS}^2 + R_P^2} \tag{12}$$

$$Q_2 = \omega \cdot R_{PS} \cdot \tau_{contact} \cdot 4\pi R_{PS}\sqrt{R_{PS}^2 + R_P^2} \tag{13}$$

$$Q_2 = 4\omega \cdot \pi R_{PS}^2 \cdot \tau_{contact} \cdot \sqrt{R_{PS}^2 + R_P^2} \tag{14}$$

3.3. Heat Generation from the Pin Tip Surface

The following equations give the heat generated from the pin tip. Assuming that the flat pin tip gives the heat generated from the pin tip, Q_3, thus:

$$dQ_3 = \omega \cdot r \cdot \tau_{contact} \cdot r d\theta dr \tag{15}$$

$$Q_3 = \int_0^{2\pi} \int_0^{R_p} \omega \cdot r \cdot \tau_{contact} \cdot r d\theta dr = \frac{2}{3}\pi\omega \cdot \tau_{contact} \cdot R_p^3 \tag{16}$$

From Equations (6), (14) and (16), Q_{Total} can be calculated as written in Equation (17),

$$Q_{Total} = Q_1 + Q_2 + Q_3 \tag{17}$$

$$Q_{Total} = \left(2\pi\omega \cdot \tau_{contact} \frac{(R_s^3 - R_p^3)}{3cos\alpha}\right) + \left(4\omega \cdot \pi R_{PS}^2 \cdot \tau_{contact} \cdot \sqrt{R_{PS}^2 + R_P^2}\right) + \left(\frac{2}{3}\pi\omega \cdot \tau_{contact} \cdot R_p^3\right) \tag{18}$$

$$Q_{Total} = \frac{2}{3}\pi\omega \cdot \tau_{contact} \cdot \left[\left(\frac{\left(R_S^3 - R_p^3\right)}{cos\alpha}\right) + \left(6R_{PS}^2\sqrt{R_{PS}^2 + R_P^2}\right) + \left(R_p^3\right)\right] \tag{19}$$

The estimated shear stress for the sliding condition is given by:

$$\tau_{contact} = p \cdot \mu \tag{20}$$

and pressure "p" is given by:

$$p = F/\pi \cdot {R_s}^2 \tag{21}$$

The energy per unit length can be calculated by dividing Equation (19) by the welding speed. Thus:

$$Q_{Energy/Length} = \frac{2\pi\omega F\mu}{3vR_S^2}\left[\left(\frac{\left(R_S^3 - R_p^3\right)}{cos\alpha}\right) + \left(6R_{PS}^2\sqrt{R_{PS}^2 + R_P^2}\right) + \left(R_p^3\right)\right] \tag{22}$$

The coefficient of friction (μ) varies with temperature [38,39]. However, in the present model, for demonstration purposes, it was considered to be 0.5. The effective energy per weld length (Q_{Eff}) [40–42] is defined as the energy per weld length multiplied by the transfer efficiency ("β" ratio of the pin length "Hp" to the workpiece thickness "t") and given by

$$Q_{Eff} = \beta \cdot Q_{Energy/Length} = \left(H_p/t\right) \times Q_{Energy/Length} \tag{23}$$

The analytical relationship developed by Hamilton et al. [40] between the temperature ratio and the effective energy level Q_{Eff} was considered. The analytical equation is given by Equation (24) as:

$$\frac{T_{max}}{T_S} = 1.56 \times 10^{-4} Q_{Eff} + 0.54 \tag{24}$$

The current model should provide an upper bound for the thermal profiles. The finite element (FE) heat flux can be related to the radial position r by using an equation developed by Khandkar et al. [41] to give

$$\dot{Q}(r) = Q_{Eff} \cdot r/\left(\frac{2}{3}\pi R_S^3 + 2\pi R_{PS}^2 H\right) \tag{25}$$

where $\dot{Q}(r)$ is the local heat flux and is linearly related to r. This FE equation has been used to calculate the moving heat input at the three different interfaces between the tool and the workpiece. Alternatively, the FE heat flux can also be related to r through tool rotation and the average shear stress to yield the same effect. The energy balance equation of the FSW process can be described by:

$$\rho c\dot{T} = k\left(\frac{\partial^2 T}{\partial x^2} + \frac{\partial^2 T}{\partial y^2} + \frac{\partial^2 T}{\partial z^2}\right) + \dot{Q} \tag{26}$$

where T is the temperature, $\dot{T}$ is the rate of change in temperature, c is the specific heat, ρ is the density, k is the thermal conductivity, and $\dot{Q}$ is the rate of moving heat generation per unit volume.

$$-k\left(\frac{\partial T}{\partial x}\right) = \alpha_{sim}(T_x - T_0) \tag{27}$$

where α_{sim} is the convection heat transfer coefficient, and its value in this paper is 15 W m^{-2} K^{-1} with an ambient temperature T_0 of 24 °C for the top and side surfaces of the workpiece.

To validate the proposed model, experimental and simulated thermal profiles are presented in this section. Figure 3a shows the experimentally measured spatial temperature distribution for the three welding speeds of 25, 50, and 75 mm/min, whereas Figure 3b shows the experimental temperature history. Good agreement is observed between the experimental and calculated values.

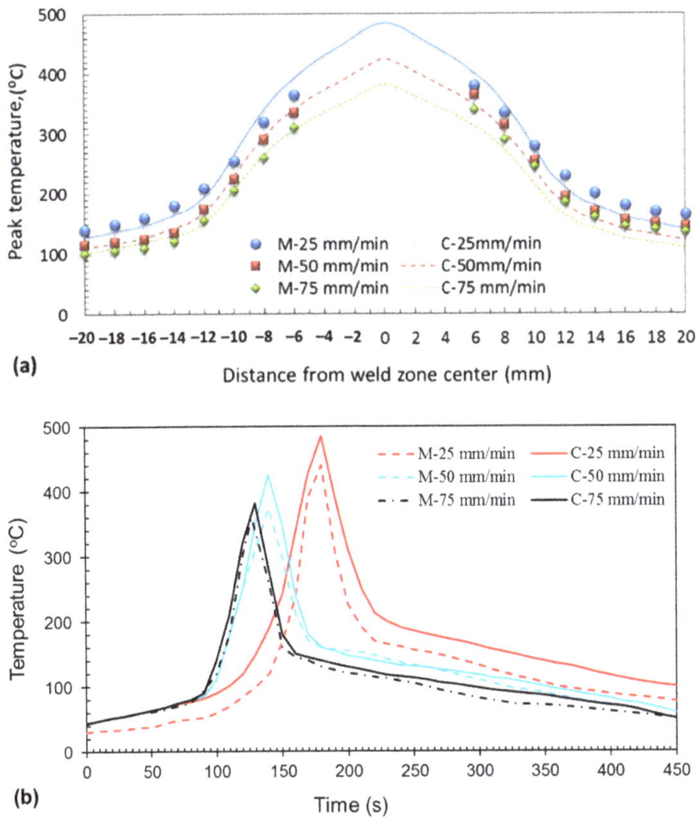

Figure 3. (**a**) Variation of calculated and experimental temperature for FSWed joints at 25, 50, and 75 mm/min welding speeds. M: refers to "measured" and C: refers to "calculated", (**a**) temperature distribution away from the weld joint center line, (**b**) The experimental and calculated temperature history.

The peak temperature at any distance from the weld zone center increased with decreasing traveling speed because of increasing heat input with decreasing traveling speed. Additionally, the maximum peak temperature was obtained at the center of the nugget zone at the three welded joints, which may be caused by the severe plastic deformation that results in a high quantity of heat input. Another dramatic observation is that a higher value of the peak temperatures was observed in the advancing side where the alloy AA5083 was placed compared with the retreating side where the alloy AA7075 was placed, and this may be attributed to the increasing rotation speed in the advancing side compared with the retreating side, causing the advancing side to be hotter [42]. The results of the proposed model indicate an increase in the heat generated at a 25 mm/min travel speed.

4. Evaluation of FSWed AA7075/AA5083 Dissimilar Joints

The results of the mechanical properties of the welded joints will be discussed with macrostructure and microstructure support. The heat generation during FSW and the temperature ratio will also be analyzed based on the proposed model and experimental measurements.

4.1. Microstructure Investigations

The macrostructure for the cross-section of the FSW 5083/7075 dissimilar joints is shown in Figure 4. A clear interface is observed in the nugget zone for all three joints. Another important observation is the superior adhesion and soundness obtained in the joints welded at travel speeds of 25 and 50 mm/min, but a tunnel defect is formed in the joint welded at a travel speed of 75 mm/min. The formation of tunnel defects has been researched by several authors [34–45], especially for friction stir welding of dissimilar aluminum alloys. The configuration of the welded joints plays an important role in the formation of defects [46] and depends mainly on the selection of the alloy to be placed on the advancing side and on the retreating side. In the present work, the weaker-strength AA5083 alloy was placed on the advancing side, and the higher-strength AA7075 alloy was placed on the retreating side. The defect was formed on the advancing side where the 5083 alloy was placed. In [47], it was concluded that defects will exist in welded joints if a higher-strength alloy is placed on the advancing side, which disagrees with the results obtained in the present work. On the other hand, since the material on the advancing side of the weld is flowing in the same direction as the tool, this can cause the material to become more fragmented and less cohesive, making it more susceptible to defect formation. Also, the material on the advancing side of the weld is subjected to more intense frictional heating than the material on the retreating side. This can cause the material to overheat and melt, making it more susceptible to defect formation regardless of the alloy type, which agrees with the results of the present work, where the defect was formed on the advancing side [48,49]. Moreover, AA5083 is a highly deformation-resistant alloy and has a high flow strength, so increasing the travel speed results in low heat input, which may cause poor intermixing between the two alloys, increasing the likelihood of defects [50]. The contribution of each alloy band was calculated approximately, as shown in Figure 4. At a travel speed of 25 mm/min, the band area of AA7075 was approximately 20.28 mm^2, higher than that of the band area of AA5083 (14.15 mm^2), whereas at a travel speed of 75 mm/min, the contribution of AA 5083 to the formation of the nugget area was higher (for AA 5083, the band area was 23.mm^2 and 11.13 mm^2 for AA 7075).

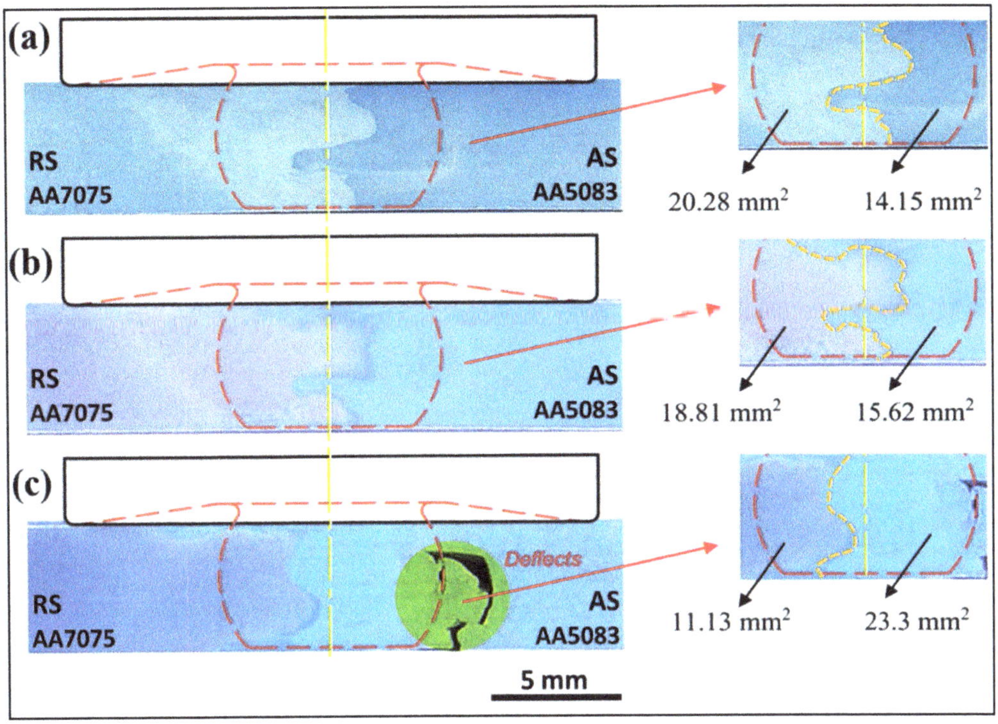

Figure 4. Macrostructure of dissimilar FSWed joints AA7075 and AA5083 at different welding speeds: (**a**) 25 mm/min, (**b**) 50 mm/min, and (**c**) 75 mm/min.

Optical microstructure investigations are paramount for FSW joints; examination of the microstructure features with grain size indicates the expected mechanical properties of welded joints. Figure 5 shows the optical microstructures and their grain size analysis of the weld zone for the dissimilar AA7075/AA5083 welded joints produced using FSW travel speeds of 25 (Figure 5a,d), 50 (Figure 5b,e), and 75 mm/min (Figure 5c,f). The microstructures of the dissimilar FSWed joints obtained fine equiaxed grains in the weld zone, as depicted in Figure 5a–c. During the FSW process, the intense localized heat generated by the rotating tool (stirring action) induces severe plastic deformation and intense stirring action in the stir zone. This action (stirring process) breaks down the coarse grains of the initial plates and promotes the formation of fine, equiaxed grains. The grain size analysis of the developed microstructures of the dissimilar welded joints revealed the significant effects of travel speeds on the grain size of the developed microstructures in the weld zone. Decreasing the travel speed from 75 to 25 mm/min tended to result in a finer grain structure within the weld zone (Figure 5d–f). The higher travel speed of 75 mm/min corresponds to a shorter exposure time to the elevated temperatures generated during the FSW process. The smallest grain size of 4.48 μm (average grain size) in the weld zone was developed in the dissimilar AA75075/AA5083 joint welded at a travel speed of 25 mm/min, while the travel speed of 75 mm/min resulted in an average grain size of 10.12 μm, as shown in Figure 5d–f. This difference in the sizes of grain structures in the weld zone is expected to affect the mechanical properties of the welded joints, such as the tensile and hardness properties.

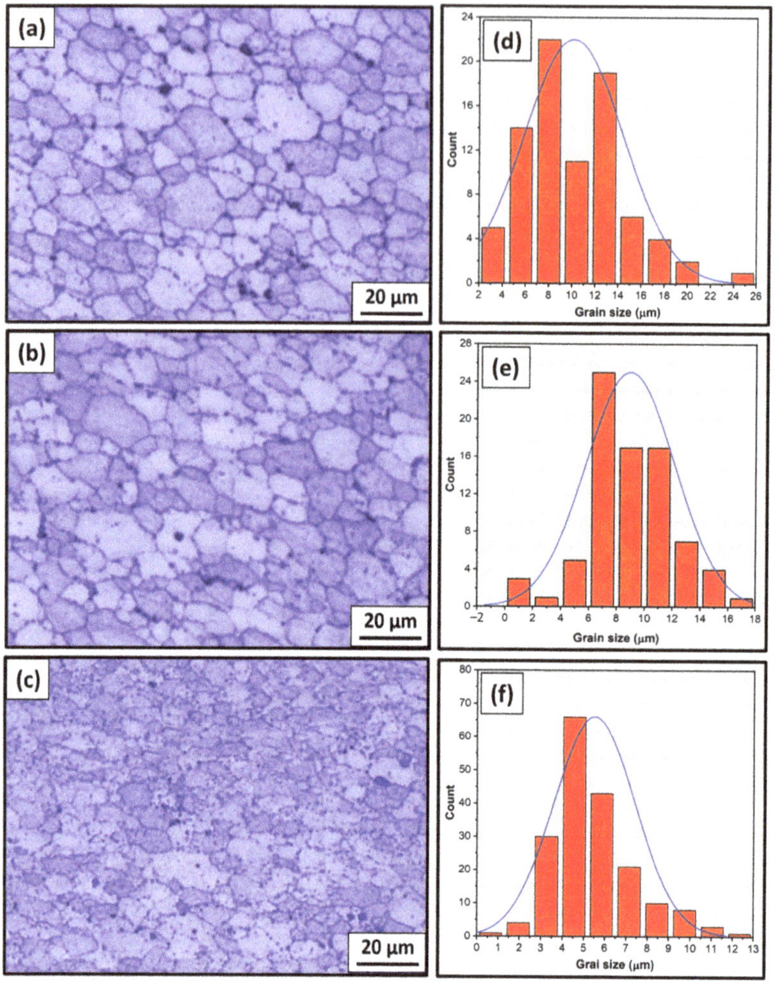

Figure 5. Microstructure and grain size analysis of dissimilar FSWed joints AA7075 and AA5083 at different welding speeds of (**a**,**d**) 25, (**b**,**e**) 50, and (**c**,**f**) 75 mm/min.

4.2. Mechanical Properties

Figure 6a illustrates stress-strain curves for joints created at welding speeds of 25, 50, and 75 mm/min, while Figure 6b illustrates the impact of travel speed on the ultimate tensile strength and joint efficiency of dissimilar FSWed butt joints welded at various travel speeds of 25, 50, and 75 mm/min at a constant rotational speed of 500 rpm (b). When the travel speed decreases, the tensile strength of the welded connection increases, since a higher heat input is possible at slower travel rates. When using a travel speed of 25 mm/min, a maximum joint efficiency of 97.7% was attained. This may be attributed to the enhancement of the bond strength [51,52], and the increase in the band area of 7075 (high strength) in the nugget zone compared with that of 5983, as shown in Figure 4. A similar work, using a different tool with an 18 mm diameter concave shoulder and a 4.8 mm long unthreaded tapered cylindrical pin, reported similar results with a maximum joint efficiency of approximately 90% at the lowest travel speed (50 mm/min) [53]. The higher weld joint efficiency in the present work compared with other similar works may be attributed to the effect of the novel tool design (hemispherical pin tool), which may result in an increase in the contact area between the tool and the joint that causes severe deformation of the stir zone, resulting in a finer structure. Additionally, some scholars used different tool designs (threaded, squared, stepped, cylindrical, and tapered) to friction stir weld dissimilar AA 5052 and AA 2024 [49]. They concluded that the maximum tensile percentage or joint efficiency of 90% is obtained using the stepped tool. More recent work [53] used a triangular threaded pin profile for FSW of dissimilar 1050-H14 and 5083-H111 aluminum alloys, and the maximum tensile percentage was 63.4%. Figure 7 shows the fracture surface of the tensile tested specimens for the dissimilar FSWed joints produced at travel speeds of (a) 25 mm/min and (b) 75 mm/min. The FSW joint produced at a 25 mm/min travel speed revealed an equiaxed deep and shallow dimple smaller than those detected for the joint welded at a 75 mm/min travel speed (Figure 7a,b). This observation confirms the higher tensile strength of the FSWed joints produced at a low travel speed of 25 mm/min than those welded at a travel speed of 75 mm/min (Figure 6).

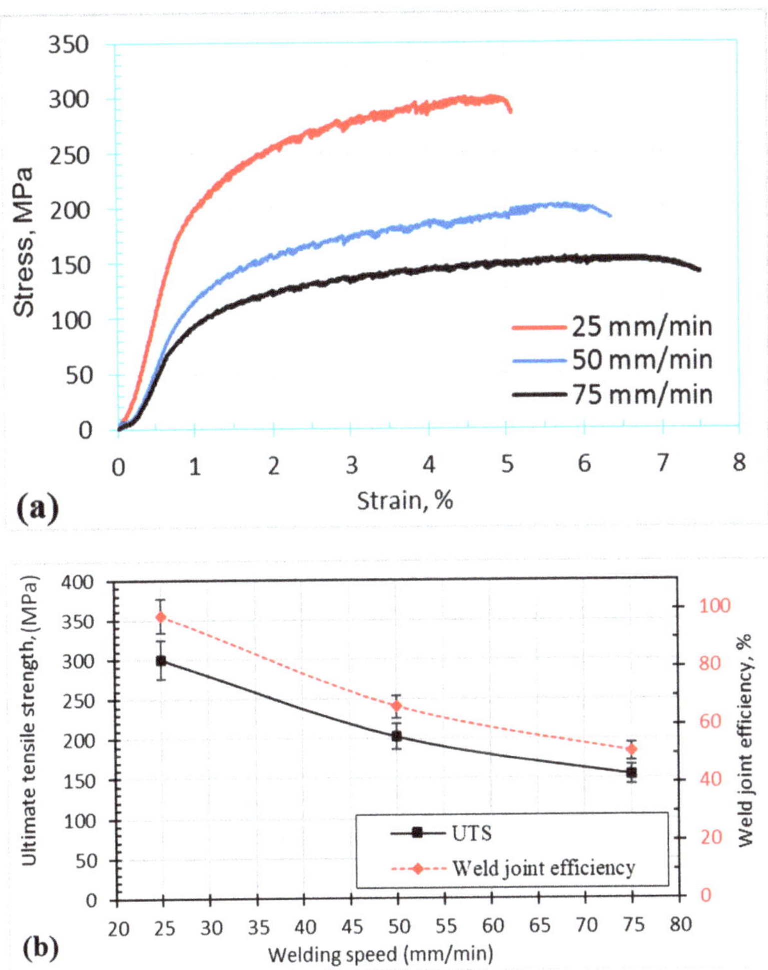

Figure 6. Tensile properties of dissimilar FSWed joints AA7075–AA5083 at different welding speeds of 25, 50, and 75 mm/min. (**a**) Stress-Strain curves; (**b**) ultimate tensile strength and welding joint efficiency.

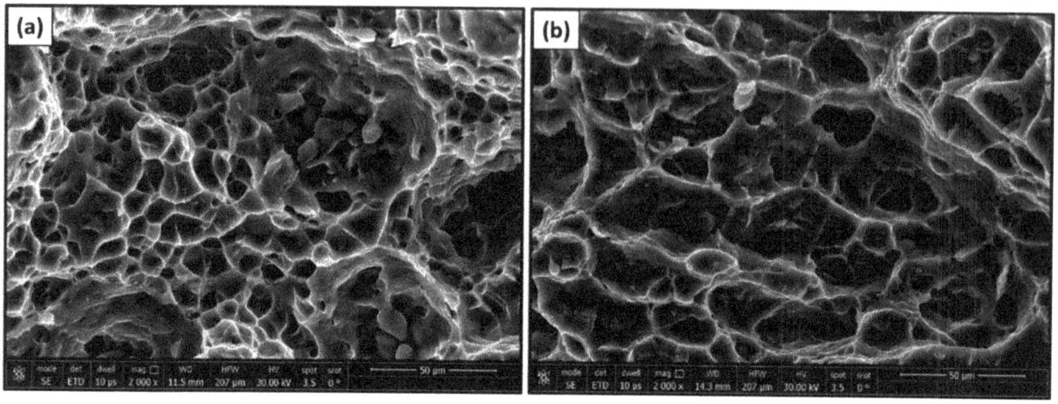

Figure 7. Fracture surface morphology of the dissimilar AA7075/AA5083 FSWed joints welded at travel speeds of (**a**) 25 mm/min and (**b**) 75 mm/min.

Figure 8 illustrates the hardness profile for the welded joints produced at different traveling speeds of 25, 50, and 75 mm/min and a constant rotation speed of 500 rpm. It can be observed that the weld zone of the dissimilar joints revealed higher hardness than the AA5083 initial materials, where the highest hardness value of 200 HV was discovered at the lowest travel speed of 25 mm/min. Furthermore, based on the hardness profile in Figure 8, it is evident that the hardness profile exhibits three distinct zones. The zone with the lowest hardness corresponds to the AA5083 side (advancing side), while the zone with the highest hardness corresponds to the AA7075 side (retreating side). The dissimilar weld zone (AA7075/AA5083) displayed hardness values that fell within the range of the other two zones, according to the experimental temperature of the dissimilar welded joints ranging from 385 to 487 °C. It has been established that the amount of heat generated depends on various factors, including the rotating speed, applied pressure, frictional conditions, and tool design. Conversely, heat dissipation is contingent upon various factors, such as the pace at which welding is conducted, the thickness of the material being welded, and the prevailing atmospheric conditions. The pace at which heat is generated and dissipated plays a crucial role in determining the extent of the temperature increase and subsequent structural modifications in the weld zone and its surrounding materials. This leads to the creation of a supersaturated solid solution within the heat-affected zones and an averaging condition in the surrounding material [47,48]. The observed changes in hardness values along the cross-section of the dissimilar joints can be attributed mostly to the structural modifications that occur.

4.3. Heat Generation and Temperature Profiles

The above-described analytical model to calculate the heat generated was utilized to better understand the impact of a hemispherical tool on the characteristics of friction stir welded connections. Figure 9a illustrates a significant link between the welding speed and heat output based on the proposed and validated models above. The relationship between the total heat produced and the energy per unit length (calculated by dividing the total heat produced by the welding speed) is very strong. It is assumed that the energy per unit length decreases when the welding speed increases. On the other hand, the total heat generated increases with increasing welding speed, which may be attributed to the increase in plunge force and torque that cause the increases in friction dissipation energy generated [50,52]. Higher tensile strength and joint efficiency were obtained at a lower welding speed (Figure 5b), which means that the higher the energy per unit length is, the stronger the weld joint becomes.

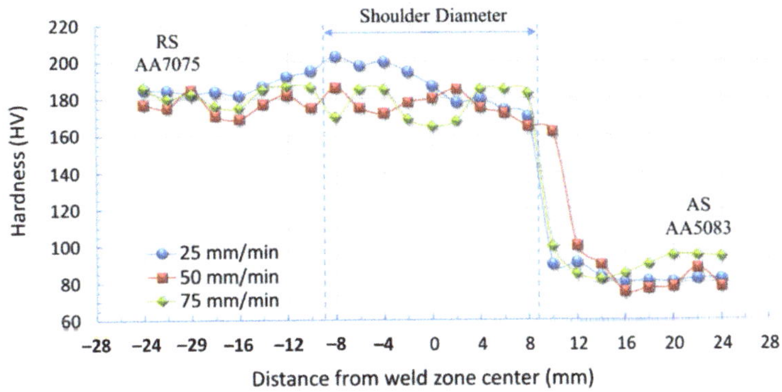

Figure 8. Hardness profiles of dissimilar FSWed joints AA7075 and AA5083.

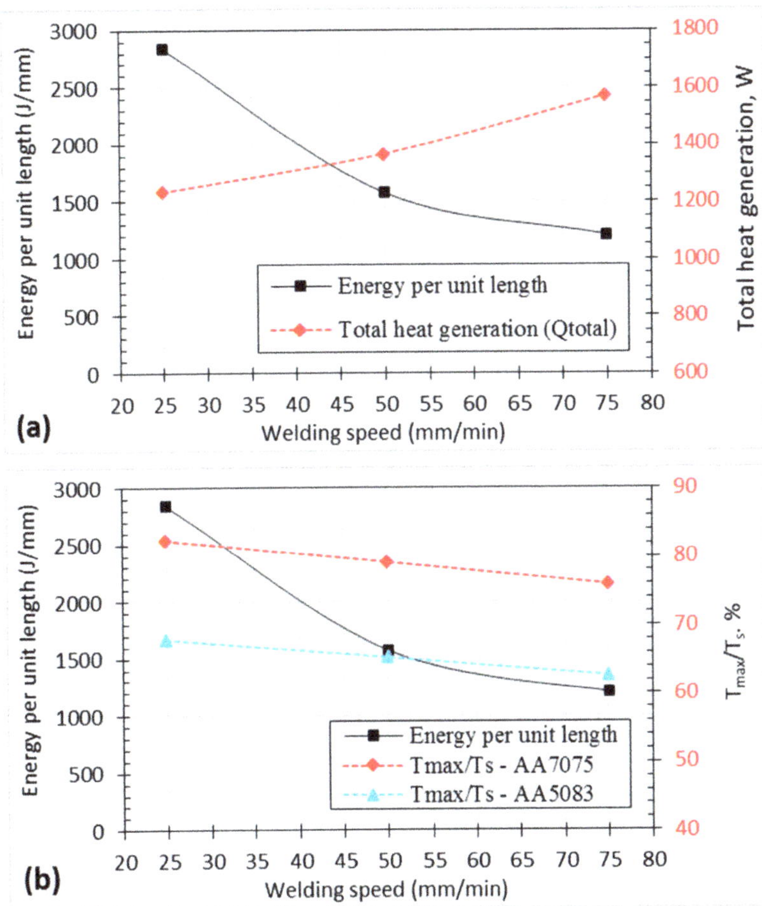

Figure 9. Effect of welding speed on the heat generation and temperature ratio for dissimilar weld joints AA 7075 and AA 5083: (**a**) effective energy per unit length and total heat generation, (**b**) effective energy per unit length and temperature ratio.

The effect of the welding speed on the temperature ratio (T_{max}/Ts), where T_{max} is the maximum temperature at the nugget zone, and Ts is the melting point for the alloy, is plotted in Figure 9b. It is clear that both energy per unit length and T_{max}/Ts ratio decrease with increasing travel speed. It is important to note that the temperature ratio for AA 7075 exceeds 75% at all traveling speeds, whereas the temperature ratio for AA5083 reaches a maximum value of 68% at a 25 mm/min travel speed and 62% at a 75 mm/min travel speed. This means that the heat input available for sound joint formation is slightly lower for the 5083 alloy than for the 7075 alloy. This may be an important concept for tunnel defect formation when the temperature ratio decreases [44,45].

5. Conclusions

A hemispherical pin tool was used for friction stir welding of dissimilar 5083 and 7075 aluminum alloys using a constant rotation speed of 500 rpm and different travel speeds of 25, 50, and 75 mm/min. An analytical model was proposed and validated experimentally. Good agreement was observed between the proposed model of heat energy generated with the associated peak temperatures and the experimentally obtained results. The energy per unit length is a more important index than the total generated heat since tensile strength and joint efficiency become optimized as energy per unit length increases. Defect-free joints were produced with travel speeds of 25 and 50 mm/min. A tunnel defect was formed on the advancing side where the 5083 alloy was placed, using a travel speed of 75 mm/min. A maximum tensile percentage of 97.7% was obtained at a travel speed of 25 mm/min. The hemispherical pin tool is suitable for friction stir welding processes and needs more attention in future work.

Author Contributions: Conceptualization, A.R.S.E., M.M.Z.A., R.I.A.E., A.A. and M.I.A.H.; methodology, A.E.E.-N., A.A.E.-A. and B.A.; software, A.R.S.E., R.I.A.E. and M.I.A.H.; validation, M.M.Z.A., A.A., A.E.E.-N., A.A.E.-A. and B.A.; formal analysis, A.R.S.E., M.M.Z.A. and A.A.; investigation, R.I.A.E., M.I.A.H., A.E.E.-N., A.A.E.-A. and B.A.; resources, A.R.S.E., M.M.Z.A., R.I.A.E., A.A. and M.I.A.H.; data curation, A.E.E.-N., A.A.E.-A. and B.A.; writing—original draft preparation, A.R.S.E., R.I.A.E., M.I.A.H. and A.A.E.-A.; writing—review and editing, M.M.Z.A. and A.A.; supervision, A.R.S.E., M.M.Z.A., R.I.A.E., A.A., M.I.A.H. and A.A.E.-A.; project administration, A.R.S.E., R.I.A.E., M.I.A.H. and A.A.E.-A.; funding acquisition, M.M.Z.A., A.A., A.A.E.-A. and B.A. All authors have read and agreed to the published version of the manuscript.

Funding: This study is supported via funding from Prince Sattam bin Abdulaziz University, project number (PSAU/2023/R/1445).

Institutional Review Board Statement: Not applicable.

Informed Consent Statement: Not applicable.

Data Availability Statement: Data will be made available upon request through the corresponding author.

Conflicts of Interest: The authors declare no conflict of interest.

References

1. Mao, Y.; Ke, L.; Liu, F.; Liu, Q.; Huang, C.; Xing, L. Effect of tool pin eccentricity on microstructure and mechanical properties in friction stir welded 7075 aluminum alloy thick plate. *Mater. Des.* **2014**, *62*, 334–343. [CrossRef]
2. Heidarzadeh, A.; Mironov, S.; Kaibyshev, R.; Çam, G.; Simar, A.; Gerlich, A.; Khodabakhshi, F.; Mostafaei, A.; Field, D.; Robson, J.; et al. Friction stir welding/processing of metals and alloys: A comprehensive review on microstructural evolution. *Prog. Mater. Sci.* **2020**, *117*, 100752. [CrossRef]
3. Çam, G.; Javaheri, V.; Heidarzadeh, A. Advances in FSW and FSSW of dissimilar Al-alloy plates. *J. Adhes. Sci. Technol.* **2023**, *37*, 162–194. [CrossRef]
4. Ipekoglu, G.; Erim, S.; Kiral, B.G.; Cam, G. Investigation inti the effect of temper condition on friction stir weldability of AA6061 Al-alloy. *Met. Mater.* **2013**, *51*, 155–163. [CrossRef]
5. Çam, G.; Güçlüer, S.; Çakan, A.; Serindag, H. Mechanical properties of friction stir butt-welded Al-5086 H32 plate. *Mater. Werkst.* **2009**, *40*, 638–642. [CrossRef]

6. Bozkurt, Y.; Salman, S.; Çam, G. Effect of welding parameters on lap shear tensile properties of dissimilar friction stir spot welded aa 5754-h22/2024-t3 joints. *Sci. Technol. Weld. Join.* **2013**, *18*, 337–345. [CrossRef]
7. İpekoğlu, G.; Gören Kıral, B.; Erim, S.; Çam, G. Investigation of the effect of temper condition on the friction-stir weldability of AA7075 Al-alloy plates. *Mater. Tehnol.* **2012**, *46*, 627–632.
8. Khalaf, H.I.; Al-Sabur, R.; Abdullah, M.E.; Kubit, A.; Derazkola, H.A. Effects of Underwater Friction Stir Welding Heat Generation on Residual Stress of AA6068-T6 Aluminum Alloy. *Materials* **2022**, *15*, 2223. [CrossRef] [PubMed]
9. Kosturek, R.; Torzewski, J.; Joska, Z.; Wachowski, M.; Śnieżek, L. The influence of tool rotation speed on the low-cycle fatigue behavior of AA2519-T62 friction stir welded butt joints. *Eng. Fail. Anal.* **2022**, *142*, 106756. [CrossRef]
10. Burek, R.; Wydrzyński, D.; Kubit, A.; Łogin, W. The influence of the shoulder depth on the properties of the thin sheet joint made by FSW technology. *Aircr. Eng. Aerosp. Technol.* **2021**, *93*, 120–126. [CrossRef]
11. Chupradit, S.; Bokov, D.O.; Suksatan, W.; Landowski, M.; Fydrych, D.; Abdullah, M.E.; Derazkola, H.A. Pin angle thermal effects on friction stir welding of AA5058 aluminum alloy: CFD simulation and experimental validation. *Materials* **2021**, *14*, 7565. [CrossRef]
12. Janeczek, A.; Tomków, J.; Fydrych, D. The Influence of Tool Shape and Process Parameters on the Mechanical Properties of AW-3004 Aluminium Alloy Friction Stir Welded Joints. *Materials* **2021**, *14*, 3244. [CrossRef]
13. Ahmed, M.M.Z.; Touileb, K.; El-Sayed Seleman, M.M.; Albaijan, I.; Habba, M.I.A. Bobbin Tool Friction Stir Welding of Aluminum: Parameters Optimization Using Taguchi Experimental Design. *Materials* **2022**, *15*, 2771. [CrossRef]
14. Ahmed, M.M.Z.; Habba, M.I.A.; Seleman, M.M.E.-S.; Hajlaoui, K.; Ataya, S.; Latief, F.H.; El-Nikhaily, A.E. Bobbin Tool Friction Stir Welding of Aluminum Thick Lap Joints: Effect of Process Parameters on Temperature Distribution and Joints' Properties. *Materials* **2021**, *14*, 4585. [CrossRef]
15. Albaijan, I.; Ahmed, M.M.Z.; Seleman, M.M.E.; Touileb, K.; Habba, M.I.A.; Fouad, R.A. Optimization of Bobbin Tool Friction Stir Processing Parameters of AA1050 Using Response Surface Methodology. *Materials* **2022**, *15*, 6886. [CrossRef]
16. Goebel, J.; Reimann, M.; Norman, A.; dos Santos, J.F. Semi-Stationary Shoulder Bobbin Tool Friction Stir Welding of AA2198-T851. *J. Mater. Process. Technol.* **2017**, *245*, 37–45. [CrossRef]
17. Sejani, D.; Li, W.; Patel, V. Stationary Shoulder Friction Stir Welding–Low Heat Input Joining Technique: A Review in Comparison with Conventional FSW and Bobbin Tool FSW. *Crit. Rev. Solid State Mater. Sci.* **2021**, *47*, 865–914. [CrossRef]
18. Shi, L.; Wu, C.S.; Liu, H.J. Numerical Analysis of Heat Generation and Temperature Field in Reverse Dual-Rotation Friction Stir Welding. *Int. J. Adv. Manuf. Technol.* **2014**, *74*, 319–334. [CrossRef]
19. Kalemba-Rec, I.; Kopyściański, M.; Miara, D.; Krasnowski, K. Effect of Process Parameters on Mechanical Properties of Friction Stir Welded Dissimilar 7075-T651 and 5083-H111 Aluminum Alloys. *Int. J. Adv. Manuf. Technol.* **2018**, *97*, 2767–2779. [CrossRef]
20. Kumar, R.; Kumar, G.; Roy, A.; Sinha, R.S.; Hasnain, S.M.M.; Prakash, O.; Ahmad, A. A Comparative Analysis of Friction Stir and Tungsten Inert Gas Dissimilar AA5082-AA7075 Butt Welds. *Mater. Sci. Energy Technol.* **2022**, *5*, 74–80. [CrossRef]
21. Kavitha, M.; Manickavasagam, V.M.; Sathish, T.; Gugulothu, B.; Sathish Kumar, A.; Karthikeyan, S.; Subbiah, R. Parameters Optimization of Dissimilar Friction Stir Welding for AA7079 and AA8050 through RSM. *Adv. Mater. Sci. Eng.* **2021**, *2021*, 9723699. [CrossRef]
22. Haribalaji, V.; Boopathi, S.; Mohammed Asif, M. Optimization of Friction Stir Welding Process to Join Dissimilar AA2014 and AA7075 Aluminum Alloys. *Mater. Today Proc.* **2022**, *50*, 2227–2234. [CrossRef]
23. Bahemmat, P.; Haghpanahi, M.; Givi, M.K.B.; Seighalani, K.R. Study on Dissimilar Friction Stir Butt Welding of AA7075-O and AA2024-T4 Considering the Manufacturing Limitation. *Int. J. Adv. Manuf. Technol.* **2012**, *59*, 939–953. [CrossRef]
24. Park, S.W.; Yoon, T.J.; Kang, C.Y. Effects of the Shoulder Diameter and Weld Pitch on the Tensile Shear Load in Friction-Stir Welding of AA6111/AA5023 Aluminum Alloys. *J. Mater. Process. Technol.* **2017**, *241*, 112–119. [CrossRef]
25. Jana, S.; Hovanski, Y.; Grant, G.J.; Mattlin, K. Effect of Tool Feature on the Joint Strength of Dissimilar Friction Stir Lap Welds. In *Friction Stir Welding and Processing VI*; John Wiley & Sons, Inc.: Hoboken, NJ, USA, 2011; pp. 205–211. [CrossRef]
26. Khajeh, R.; Javidani, M.; Mofarrehi, M.; Chen, X.-G.; Ahmed, M.; Farzaneh, A.; Heidarzadeh, A. Enhancing microstructure and mechanical properties of laser powder bed fusion-fabricated AlSi10Mg alloy through tailored friction stir processing and post-heat treatment. *Mater. Sci. Eng. A* **2024**, *889*, 145855. [CrossRef]
27. Venkateshkannan, M.; Rajkumar, V.; Sadeesh, P.; Arivazhagan, N.; Narayanan, S.; Devendranath, R.K. Influences of Tool Geometry on Metallurgical and Mechanical Properties of Friction Stir Welded Dissimilar AA 2024 and AA 5052. *Procedia Eng.* **2014**, *75*, 154–158. [CrossRef]
28. RajKumar, V.; VenkateshKannan, M.; Sadeesh, P.; Arivazhagan, N.; Devendranath Ramkumar, K. Studies on Effect of Tool Design and Welding Parameters on the Friction Stir Welding of Dissimilar Aluminium Alloys AA 5052—AA 6061. *Procedia Eng.* **2014**, *75*, 93–97. [CrossRef]
29. Reza-E-Rabby, M.; Tang, W.; Reynolds, A.P. Effect of Tool Pin Features on Process Response Variables during Friction Stir Welding of Dissimilar Aluminum Alloys. *Sci. Technol. Weld. Join.* **2015**, *20*, 425–432. [CrossRef]
30. Sued, M.K.; Pons, D.; Lavroff, J.; Wong, E.H. Design Features for Bobbin Friction Stir Welding Tools: Development of a Conceptual Model Linking the Underlying Physics to the Production Process. *Mater. Des.* **2014**, *54*, 632–643. [CrossRef]
31. Sadoun, A.M.; Wagih, A.; Fathy, A.; Essa, A.R.S. Effect of Tool Pin Side Area Ratio on Temperature Distribution in Friction Stir Welding. *Results Phys.* **2019**, *15*, 102814. [CrossRef]

32. Kalemba-Rec, I.; Hamilton, C.; Kopyściański, M.; Miara, D.; Krasnowski, K. Microstructure and Mechanical Properties of Friction Stir Welded 5083 and 7075 Aluminum Alloys. *J. Mater. Eng. Perform.* **2017**, *26*, 1032–1043. [CrossRef]
33. Heurtier, P.; Jones, M.J.; Desrayaud, C.; Driver, J.H.; Montheillet, F.; Allehaux, D. Mechanical and Thermal Modelling of Friction Stir Welding. *J. Mater. Process. Technol.* **2006**, *171*, 348–357. [CrossRef]
34. Ahmed, M.M.Z.; Habba, M.I.A.; Jouini, N.; Alzahrani, B.; El-Sayed Seleman, M.M.; El-Nikhaily, A. Bobbin Tool Friction Stir Welding of Aluminum Using Different Tool Pin Geometries: Mathematical Models for the Heat Generation. *Metals* **2021**, *11*, 438. [CrossRef]
35. Essa, A.R.S.; Ahmed, M.M.Z.; Aboud, A.R.K.; Alyamani, R.; Sebaey, T.A. Prediction of Tool Eccentricity Effects on the Mechanical Properties of Friction Stir Welded AA5754-H24 Aluminum Alloy Using ANN Model. *Materials* **2023**, *16*, 3777. [CrossRef]
36. Schmidt, H.; Hattel, J.; Wert, J. An Analytical Model for the Heat Generation in Friction Stir Welding. *Model. Simul. Mater. Sci. Eng.* **2004**, *12*, 143–157. [CrossRef]
37. Mishra, R.S. *Friction Stir Welding and Processing*; ASM International: Materials Park, OH, USA, 2007; ISBN 978-0-87170-840-3.
38. Mohamed, A.; Mohammed, M.M.; Ibrahim, A.F.; El-Kady, O.A. Effect of Nano Al2O3 Coated Ag Reinforced Cu Matrix Nanocomposites on Mechanical and Tribological Behavior Synthesis by P/M Technique. *J. Compos. Mater.* **2020**, *54*, 4921–4928. [CrossRef]
39. Nataliia, D.; Erik, G.; Igor, Z.; Klaus, Z. Mathematical Modeling of Friction Stir Welding Considering Dry and Viscous Friction. *Appl. Math. Model.* **2019**, *67*, 1–8. [CrossRef]
40. Hamilton, C.; Dymek, S.; Sommers, A. A Thermal Model of Friction Stir Welding in Aluminum Alloys. *Int. J. Mach. Tools Manuf.* **2008**, *48*, 1120–1130. [CrossRef]
41. Khandkar, M.Z.H.; Khan, J.A.; Reynolds, A.P. Prediction of Temperature Distribution and Thermal History during Friction Stir Welding: Input Torque Based Model. *Sci. Technol. Weld. Join.* **2003**, *8*, 165–174. [CrossRef]
42. Ajri, A.; Shin, Y.C. Investigation on the Effects of Process Parameters on Defect Formation in Friction Stir Welded Samples via Predictive Numerical Modeling and Experiments. *J. Manuf. Sci. Eng. Trans. ASME* **2017**, *139*, 111009. [CrossRef]
43. Morisada, Y.; Imaizumi, T.; Fujii, H. Clarification of Material Flow and Defect Formation during Friction Stir Welding. *Sci. Technol. Weld. Join.* **2015**, *20*, 130–137. [CrossRef]
44. Mehta, K.P.; Badheka, V.J. Effects of Tool Pin Design on Formation of Defects in Dissimilar Friction Stir Welding. *Procedia Technol.* **2016**, *23*, 513–518. [CrossRef]
45. Mastanaiah, P.; Sharma, A.; Reddy, G.M. Dissimilar Friction Stir Welds in AA2219-AA5083 Aluminium Alloys: Effect of Process Parameters on Material Inter-Mixing, Defect Formation, and Mechanical Properties. *Trans. Indian Inst. Met.* **2016**, *69*, 1397–1415. [CrossRef]
46. Ghosh, M.; Kumar, K.; Kailas, S.V.; Ray, A.K. Optimization of Friction Stir Welding Parameters for Dissimilar Aluminum Alloys. *Mater. Des.* **2010**, *31*, 3033–3037. [CrossRef]
47. Ahmed, M.S.I.; Ahmed, M.M.Z.; Abd El-Aziz, H.M.; Habba, M.I.A.; Ismael, A.F.; El-Sayed Seleman, M.M.; Abd El-Aty, A.; Alamry, A.; Alzahrani, B.; Touileb, K.; et al. Cladding of Carbon Steel with Stainless Steel Using Friction Stir Welding: Effect of Process Parameters on Microstructure and Mechanical Properties. *Crystals* **2023**, *13*, 1559. [CrossRef]
48. Msomi, V.; Mbana, N.; Mabuwa, S. Microstructural Analysis of the Friction Stir Welded 1050-H14 and 5083-H111 Aluminium Alloys. *Mater. Today Proc.* **2019**, *26*, 189–192. [CrossRef]
49. Mishra, R.S.; Ma, Z.Y. Friction Stir Welding and Processing. *Mater. Sci. Eng. R Rep.* **2005**, *50*, 1–78. [CrossRef]
50. Aziz, S.B.; Dewan, M.W.; Huggett, D.J.; Wahab, M.A.; Okeil, A.M.; Liao, T.W. Impact of Friction Stir Welding (FSW) Process Parameters on Thermal Modeling and Heat Generation of Aluminum Alloy Joints. *Acta Met. Sin. (Engl. Lett.)* **2016**, *29*, 869–883. [CrossRef]
51. Ahmed, M.M.Z.; Seleman, M.M.E.-S.; Fydrych, D.; Çam, G. Review on Friction Stir Welding of Dissimilar Magnesium and Aluminum Alloys: Scientometric Analysis and Strategies for Achieving High-Quality Joints. *J. Magnes. Alloy.* **2023**, *11*, 4082–4127. [CrossRef]
52. El-Aty, A.A.; Xu, Y.; Xie, W.; Xia, L.-L.; Hou, Y.; Zhang, S.; Ahmed, M.M.Z.; Alzahrani, B.; Ali, A.; Huang, X.; et al. Finite Element Analysis and Experimental Study of Manufacturing Thin-Walled Five-Branched AISI 304 Stainless Steel Tubes with Different Diameters Using a Hydroforming Process. *Materials* **2024**, *17*, 104. [CrossRef]
53. Ahmed, M.M.Z.; El-Sayed Seleman, M.M.; Fydrych, D.; Çam, G. Friction Stir Welding of Aluminum in the Aerospace Industry: The Current Progress and State-of-the-Art Review. *Materials* **2023**, *16*, 2971. [CrossRef] [PubMed]

materials

MDPI

Article

Inverse Neural Network Approach for Optimizing Chemical Composition in Shielded Metal Arc Weld Metals

Taehyun Yoon [1], Young IL Park [1], Jaewoong Kim [2,*] and Jeong-Hwan Kim [1,*]

[1] Department of Naval Architecture and Offshore Engineering, Dong-A University, Busan 49315, Republic of Korea; yoon.th2718@gmail.com (T.Y.); parkyi1973@dau.ac.kr (Y.I.P.)
[2] Gwangju Ppuri Technology Support Center, Korea Institute of Industrial Technology, Gwangju 11109, Republic of Korea
* Correspondence: kjw0607@kitech.re.kr (J.K.); jhkim81@dau.ac.kr (J.-H.K.)

Abstract: This study presents a hybrid machine learning framework combining an artificial neural network and a genetic algorithm to optimize chemical compositions of shielded metal arc weld metals for achieving targeted mechanical properties. First, a neural network model was trained using a large experimental database provided by Dr. Glyn M. Evans, which includes the chemical compositions and mechanical properties of over 950 shielded metal arc weld metals. The neural network model, optimized via Bayesian optimization, demonstrated high predictive accuracy for properties such as yield strength, ultimate tensile strength, and Charpy impact transition temperatures. To enable inverse design, a genetic algorithm-based optimization was applied to the trained neural network model, iteratively exploring the composition space to find optimal elemental combinations that match predefined mechanical property targets. The proposed hybrid approach successfully identified multiple feasible compositions that closely match the desired mechanical behavior, demonstrating the potential of neural network-assisted inverse design in welding alloy development.

Keywords: inverse neural network; shielded metal arc weld; genetic algorithm; chemical composition

1. Introduction

The mechanical properties of welds are significantly influenced by their chemical composition, which has led to extensive research to understand this relationship. Mechanical tests have been conducted on various materials and welding methods, resulting in the development of a comprehensive database that links chemical composition to the mechanical properties of welds [1–5]. In May 2015, Dr. Glyn M. Evans published an extensive database on ResearchGate, encompassing over 950 shielded metal arc (SMA) weld metal compositions. Each composition includes information on 16 elements, including iron (Fe), and six mechanical properties, such as yield strength and tensile strength [5].

Based on this database, numerous studies have analyzed the influence of individual alloying elements on the mechanical properties of welds [6–10]. Traditional statistical techniques such as constraint-based models, multiple regression analysis, and cluster analysis have been employed to explore these relationships. However, the complex interactions between the variables make it challenging to fully untangle these relationships using traditional methods alone.

Artificial neural networks (ANNs) have emerged as a promising alternative to address these issues [11,12]. ANNs are particularly useful in scenarios involving many inputs, outputs, and nonlinear relationships, as they can efficiently perform regression analysis without requiring prior assumptions about the relationships between variables. Consequently, ANNs are effective in predicting the mechanical properties of welds based on their chemical composition. For instance, ANN-based prediction models have been developed for various materials and welding conditions, demonstrating their applicability. Park et al. [13] developed an ANN-based model to predict the yield strength of austenitic stainless steel welds. Sampath [10] proposed an ANN model to predict the Charpy V-notch impact toughness of high-strength steel weld metals based on Evans's database. Bera and Das [14] developed an ANN model to predict the ultimate tensile strength (UTS), elongation, and Rockwell hardness on the B scale (HRB) for gas metal arc welding (GMAW) of dissimilar steels using current, voltage, and gas pressure as inputs. Payares-Asprino [15] presented an ANN model to predict the yield strength, tensile strength, elongation, and fracture location of duplex stainless steel (SAF 2205) welds in a robotic GMAW process under varying welding conditions. Jung et al. [16] developed an artificial intelligence (AI)-based model to predict the tensile properties of high-strength steels using microstructural factors and chemical compositions. Mezher et al. (2024a) applied various ANN architectures to predict the quality of resistance spot welding (RSW) for AISI 304 stainless steel and quantitatively analyzed the changes in shear strength and nugget diameter according to process parameters [17]. In a subsequent study, Mezher et al. (2024b) investigated dissimilar metal RSW between titanium alloy and AISI 304 austenitic stainless steel using not only ANNs but also Random Forest and CatBoost algorithms and quantitatively assessed the relative importance of input variables [18].

Despite such advantages, ANN models face challenges such as vanishing gradients and overfitting. To overcome these issues, researchers have explored new activation functions and optimization techniques. In addition, recent advancements in deeper networks and effective optimization methods have significantly enhanced model performance. These improvements have enabled the development of robust models that can avoid overfitting and achieve optimal prediction accuracy [19–28].

First, this study developed a model to predict the mechanical properties of welds based on their chemical compositions, utilizing the experimental database of SMA weld metals published by Dr. Glyn M. Evans [5] and implementing a multilayer ANN structure. Bayesian optimization techniques were used to determine the optimal number of layers and nodes and establish when to stop training. The objective was to create a model that can efficiently predict six mechanical properties of welds using only their chemical compositions.

Next, this research introduced an inverse neural network approach. While traditional methods focus on predicting mechanical properties based on chemical compositions, inverse neural networks can identify combinations of chemical components that achieve specific mechanical properties. This capability is particularly valuable to practitioners and developers who seek to derive the optimal chemical composition to achieve desired mechanical characteristics. To facilitate this, this paper utilized a genetic algorithm (GA) to explore the optimal chemical composition. GAs are powerful tools for discovering optimal combinations by mimicking the principles of natural selection. They also excel at exploring global optima through repeated selection, crossover, and mutation by mimicking the theory of natural selection and evolution. These properties make them well-suited for optimizing complex objective functions defined in high-dimensional spaces, such as those encountered in the chemical composition design problem addressed in this study.

In recent years, hybrid optimization approaches that combine machine learning (ML) techniques with GAs have garnered a significant amount of attention. These combined methods effectively address the limitations of single algorithms and enhance both predictive and exploratory performance. ML-based predictive models can quickly estimate performance within a design space, while GAs help identify combinations that meet specific target conditions through search-based optimization. The integration of these two techniques has been actively applied across various fields, particularly in materials science, where their effectiveness has been demonstrated in alloy composition design. Lee et al. [29] used an ML-aided GA approach to efficiently explore the optimal composition and processing conditions for medium-Mn steel with improved tensile strength and elongation. By integrating data-driven prediction models with GA-based inverse design techniques, ultra-high-strength compositions were successfully obtained. In addition, the effects of trace alloying elements such as Ti, V, and Mo on mechanical properties were investigated through microstructural analysis, demonstrating how an ML–GA-based framework can advance alloy design. Bhat et al. [30] combined GAs with class-based models to simultaneously optimize the strength and elongation of aluminum alloys. The researchers classified different aluminum alloy classes, trained individual regression models for each class, and linked these models with GAs to explore optimal compositions and processing conditions specific to each. Schaufelberger et al. [31] introduced an uncertainty-controlled GA based on ensemble MA predictions to explore the high-dimensional chemical space for designing singlet fission materials. Improving the synergy between ML techniques and GAs is becoming an essential approach to solving inverse design problems [29–41].

This study proposes a hybrid GA–ANN-based inverse design prediction model that overcomes the limitations of conventional forward-only ANN models by enabling goal-oriented exploration of chemical compositions based on target mechanical properties. The model integrates a high-accuracy ANN, trained on a large-scale weld metal dataset, with a genetic algorithm to iteratively search for optimal compositions that minimize the error between predicted and desired properties. By capturing the complex nonlinear relationships between composition and properties, and automating the optimization process, the framework reduces the need for extensive physical testing, lowers development costs, and improves design efficiency. Repeated validations confirmed its stable and reliable performance, demonstrating potential for broad applicability across various welding processes and alloy design tasks.

This paper is structured as follows: First, it discusses the relevant database and the basic statistical properties of the data. Second, the development of an ANN model for predicting mechanical properties is detailed. Finally, it presents the development of a model utilizing an inverse neural network approach and discusses the results.

2. Database

2.1. Specifications of the Database

The relationship between chemical composition and mechanical properties of welds was investigated based on a large-scale database of SMA weld metals published by Dr. Glyn M. Evans [5]. Table 1 presents the descriptive statistics of the data, which consists of the chemical composition of 16 elements and their corresponding six mechanical properties. The mechanical properties include yield strength (YS), ultimate tensile strength (UTS), elongation (El), reduction of area (RA), and transition temperature at 100 J and 28 J in the Charpy V-notch test (Temp-100J, Temp-28J). The presented statistical indicators provide fundamental insights into the distribution trends and variability among the characteristics. In particular, descriptive statistics such as the minimum, maximum, mean, and standard deviation of each variable enable a clear understanding of the range and scale of both chemical compositions and mechanical properties. This information is valuable for establishing effective data preprocessing and analysis strategies for the input variables.

Table 1. Summary of chemical composition and mechanical properties.

Element	Min. Value	Max. Value	Mean Value	Std. Dev.	Unit
C	0.035	0.152	0.071	0.011	wt%
Mn	0.23	2.1	1.262	0.393	wt%
Si	0.001	1.11	0.352	0.104	wt%
S	0.003	0.046	0.007	0.002	wt%
P	0.003	0.04	0.008	0.003	wt%
Ti	1	770	115.541	152.679	ppm
B	1	200	21.387	44.732	ppm
Al	1	680	40.307	111.877	ppm
N	35	270	92.916	47.832	ppm
O	217	1535	397.230	87.849	ppm
Cr	0.026	3.5	0.138	0.447	wt%
Ni	0.03	5.48	0.295	0.982	wt%
Mo	0.005	1.16	0.050	0.201	wt%
V	3	2873	48.924	237.194	ppm
Cu	0.02	2.04	0.071	0.235	wt%
Nb	3	980	19.046	84.891	ppm
YS	310	1026	505.589	77.196	N/mm^2
UTS	345	1123	578.349	77.657	N/mm^2
El	7.4	35.8	26.009	3.783	%
RA	10.8	87.8	75.719	4.608	%
Temp-100J	−89	45	−43.265	23.500	°C
Temp-28J	−145	77	−69.162	23.956	°C

2.2. Correlation Analysis

To assess the approximate correlation between each dataset, Pearson correlation analysis was performed using Equation (1) below, where r is the Pearson correlation coefficient, and x and y are the target variables being analyzed. Figure 1 visually represents the Pearson correlation coefficients between the variables. A value close to 1 indicates a strong positive correlation, while a value close to -1 indicates a strong negative correlation.

$$r = \frac{\sum (x_i - \overline{x})(y_i - \overline{y})}{\sqrt{\sum (x_i - \overline{x})^2 \sum (y_i - \overline{y})^2}} \quad (1)$$

The analysis found that none of the chemical compositions had a correlation coefficient greater than 0.5 with any of the mechanical properties. This outcome is due to the limitations of Pearson correlation analysis, which only examines linear relationships between two variables and is not effective in analyzing nonlinear interactions among chemical composition ratios or between mechanical properties. Specifically, Pearson correlation analysis fails to account for multivariate nonlinear interactions among elemental compositions, multivariate nonlinear interactions between mechanical properties, and multivariate nonlinear interactions between composition ratios and physical properties. To overcome these limitations, this study employs an ANN to analyze multivariate nonlinear interactions.

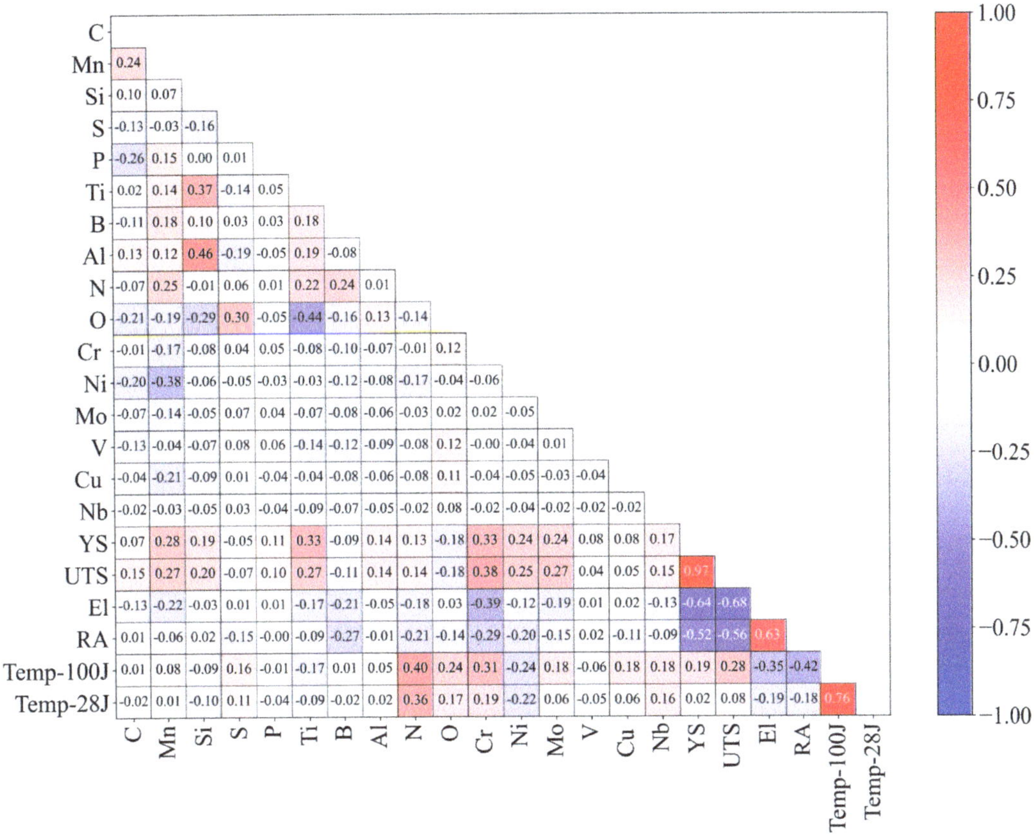

Figure 1. Pearson correlation matrix of chemical composition and mechanical properties.

3. Methodology

The hybrid genetic algorithm–artificial neural network (hybrid GA–ANN) applied in this study consists of an ANN prediction model and a GA evaluation model, which combines the powerful predictive capabilities of ANNs with the optimization ability of GAs to facilitate prediction and inverse operations. The overall process involves two main steps, as shown in Figure 2. Step 1 consists of training the ANN, which is used to build a data-driven predictive model. Step 2 involves the process of performing inverse operations based on the trained ANN model to derive the optimal elemental composition ratio that satisfies the target mechanical properties.

In Step 1, the ANN is trained using SMA weld metal data. The elemental composition ratio is set as the input, while the corresponding mechanical properties serve as the output. The model's hyperparameters, including the number of nodes per hidden layer, the number of hidden layers, and the number of epochs, which are important factors that determine the structure and performance of the neural network, are optimized using Bayesian optimization. The ANN is then trained based on these optimized hyperparameters. The performance of the trained model is evaluated using statistical metrics such as mean squared error (MSE) and the coefficient of determination (R^2) and finally stored.

In Step 2, after setting and inputting the initial values of the target mechanical properties and elemental compositions, the ANN trained in Step 1 is combined with a GA to identify the combination of elemental compositions that satisfy the target mechanical properties. First, the GA evaluation model predicts the mechanical properties based on the input values of the elemental compositions using the trained ANN. The predicted mechanical properties are compared to the target values, resulting in the elimination of unsuitable elemental compositions while suitable ones are passed on to the next generation. GAs are search algorithms designed to find data adaptable to their environment. Compositions that do not meet the evaluation criteria are discarded, while those that do are optimized through successive generations, retaining their genetic information. During this process, mutation operations are applied to generate new combinations of composition ratios. During mutation, small random changes are introduced to the existing compositions, helping to expand the search space and avoid local optima. Eventually, this process leads to a composition that satisfies the desired mechanical properties.

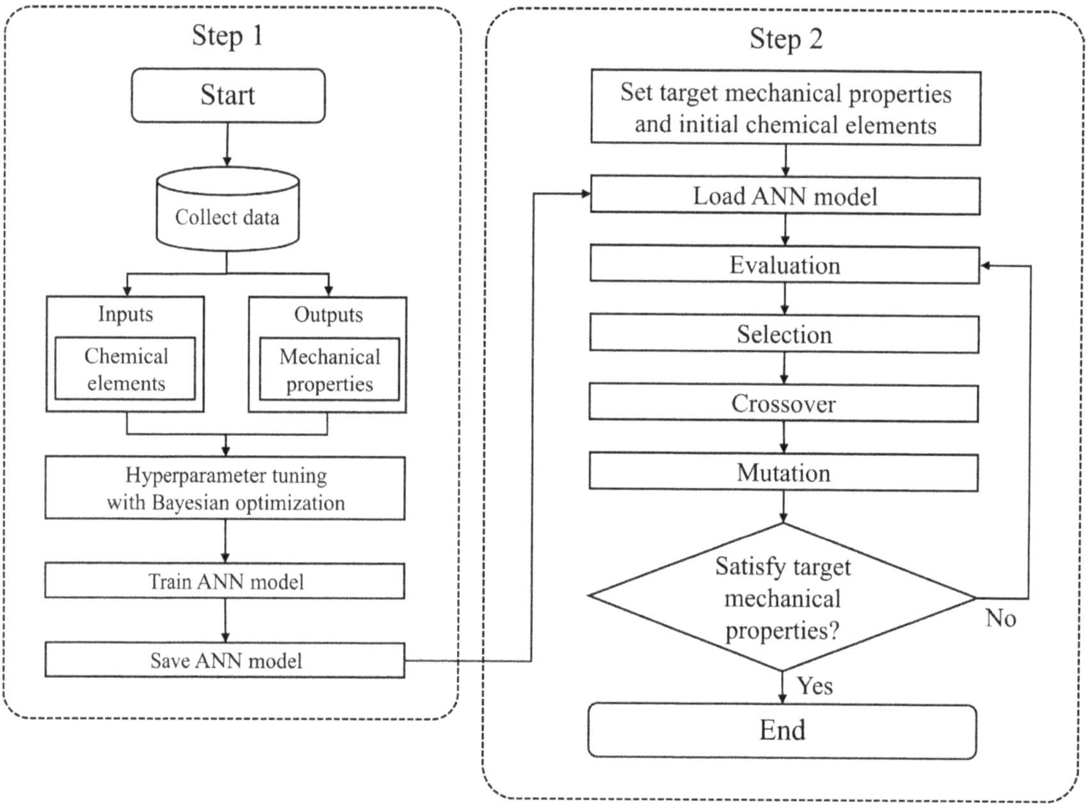

Figure 2. Workflow of the hybrid GA–ANN process.

Figure 3 illustrates the overall workflow of the GA. On the left, a representation of a chromosome shows how chromosomes are structured within the population. On the right, a flowchart describes the main steps of the GA: initialization, evaluation, selection, crossover, and mutation.

Figure 3a depicts the initial stage of the GA. The initial population represents potential solutions in the form of chromosomes. During the evaluation phase, a fitness evaluation determines how suitable each chromosome is for solving the problem. The selection phase then selects the chromosomes that will be passed on to the next generation, where individuals with higher fitness are more likely to be selected.

Figure 3b illustrates how the crossover operation is performed. Selected parent chromosomes exchange genetic material to create new offspring, which increases population diversity and helps expand the search space.

Figure 3c shows the mutation process. In this stage, specific gene materials of certain chromosomes are randomly altered to avoid local optima and facilitate more effective exploration of the search space. Mutation enhances the chances of discovering new optimal solutions, making the overall search process more flexible.

Finally, the GA iterates through the following steps: evaluation, selection, crossover, and mutation. This process continues until the predefined termination criteria are met, which may include reaching an optimal fitness value or exceeding a maximum number of generations. If the termination condition is not satisfied, the process repeats until a satisfactory solution is found.

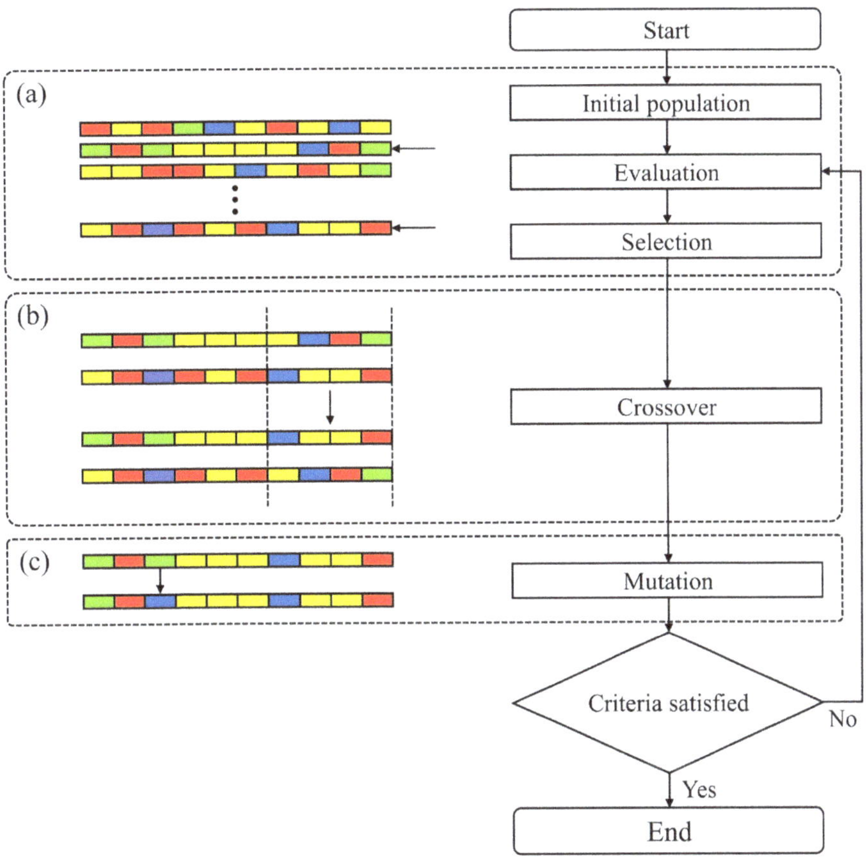

Figure 3. Workflow of the genetic algorithm process.

4. Machine Learning Model

4.1. Hyperparameter Optimization

The Bayesian optimization method was applied to identify the optimal hyperparameters. This technique utilizes a probability model based on Bayes's theorem to efficiently explore hyperparameters [19,20]. Bayesian optimization seeks the x^+ that maximizes the objective function f and is defined as shown in Equation (2), where A is the search range.

$$x^+ = arg \max_{x \in A} f(x) \tag{2}$$

The core idea of Bayesian optimization is expressed in Equation (3). It shows the relationship between the prior probability $P(M)$ of an existing model M and the probability $P(E|M)$ that E will be observed in the model when evidence data E is observed. This process involves updating the posterior probability $P(M|E)$ during optimization.

$$P(M|E) \propto P(E|M)P(M) \tag{3}$$

In executing Bayesian optimization, a stochastic model based on a Gaussian process approach is utilized to approximate f and determine the optimal sampling point. This approach balances exploration (testing unknown regions) and exploitation (exploring regions based on the optimal values found in the existing sampling process). The neural network was trained using the Adam (Adaptive Moment Estimation) optimizer, which adaptively adjusts the learning rate for each parameter. The operating principle of this optimization method is described in detail in Section 4.2. The final hyperparameters determined through this process are summarized in Table 2.

Table 2. Optimized hyperparameters.

Hyperparameters	Values
Number of nodes per hidden layer	423
Number of hidden layers	4
Epochs	859

4.2. Model Structure

This section describes the structure of the ANN. An ANN consists of an input layer, one or more hidden layers, and an output layer, based on a multilayer perceptron structure.

Table 1 shows the data provided to the input and output layers of the ANN. The input data comprises the 16 elemental composition ratios of SMA weld metals, including elements such as carbon (C), manganese (Mn), and silicon (Si). These elements serve as the independent variables the ANN is trained to use for predictions.

The output layer receives the data that the ANN aims to predict, which includes six mechanical properties of the SMA weld metals, as detailed in Table 1. These properties include yield strength (YS), ultimate tensile strength (UTS), elongation (El), reduction of area (RA), and transition temperature at 100 J and 28 J in the Charpy V-notch test (Temp-100J, Temp-28J). These are the dependent variables that the ANN needs to predict.

Figure 4 illustrates the overall structure of the ANN. To optimize model performance, Bayesian optimization was applied to explore the hyperparameters. The resulting optimal model features four hidden layers, each containing 423 nodes.

In the neural network, the Rectified Linear Unit (ReLU) was used as the activation function. The ReLU function outputs 0 when the input is less than or equal to 0 and returns the input value itself when the input is greater than 0. It is defined as follows in Equation (4):

$$\text{ReLU} = \max(0, x) \tag{4}$$

Compared to traditional activation functions such as sigmoid and tanh, ReLU has a simpler computational structure and is effective in alleviating the vanishing gradient problem during training, making it advantageous for stable learning in deep neural networks. For these reasons, ReLU was applied not only to the hidden layers but also to the output layer in this study.

The model was trained using the Adam (Adaptive Moment Estimation) optimizer. Adam is based on stochastic gradient descent (SGD) and estimates both the first moment (mean of the gradients) and the second moment (mean of the squared gradients) for each parameter during training. By automatically adjusting the learning rate for each parameter, Adam suppresses excessive oscillations and facilitates fast and stable convergence.

During the training process, mean squared error (MSE) was used as the loss function. MSE is defined as in Equation (5), where n represents the number of data points, y_i is the actual value to be predicted, and $\hat{y}_i$ is the value predicted by the ANN. This loss function evaluates errors by measuring the mean squared difference between the predicted and actual values.

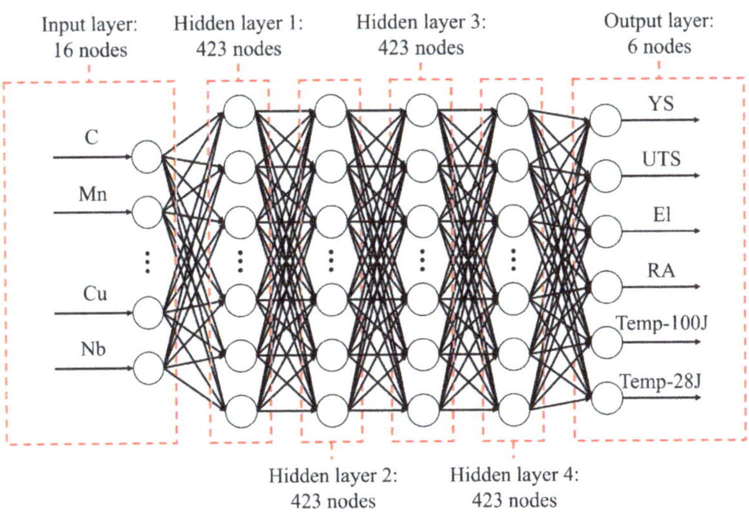

Figure 4. Structure of the prediction model.

Figure 5 shows the MSE progression throughout the training process. The horizontal axis represents the training epochs, and the vertical axis represents the MSE values, with the solid red line indicating the training error and the dashed blue line representing the validation error. As shown in Figure 5, the loss value decreases sharply during the initial training phase and then gradually converges, indicating that the model is moving toward an optimal state and that the training process is stable. Additionally, the training and validation errors remain at similar levels, suggesting that overfitting has not occurred.

$$MSE = \frac{1}{n}\sum_{i=1}^{n}(y_i - \hat{y}_i)^2 \quad (5)$$

4.3. *Performance of Machine Learning Model*

The performance of the prediction model was evaluated using data that was not included in the training set. Figure 6 displays the performance of the ANN prediction model. The results indicate very high prediction accuracy for yield strength and ultimate tensile strength. The predictions for Temp-100J and Temp-28J show relatively lower accuracy compared to YS and UTS yet still maintain a high level of precision. In contrast, El and RA exhibit large scatters in the predicted data and relatively high errors.

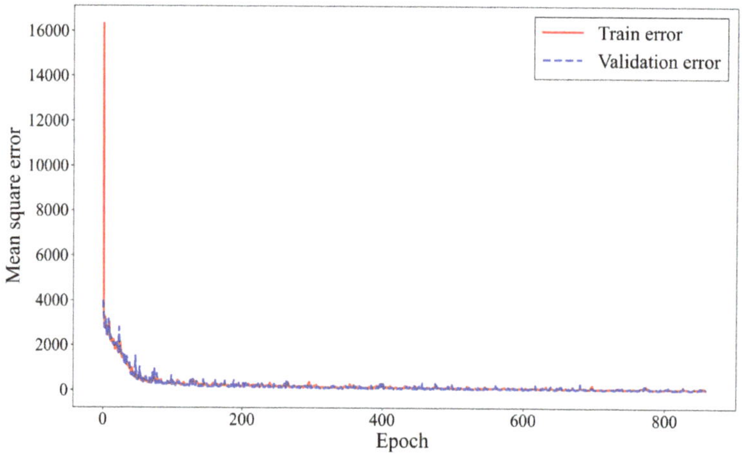

Figure 5. Prediction model performance during training.

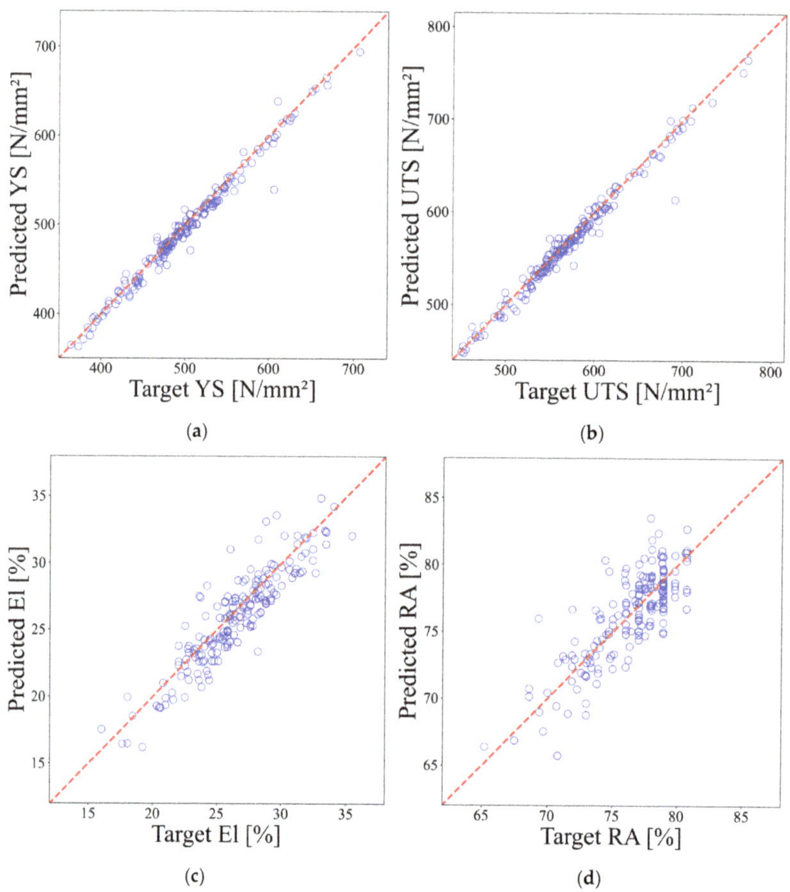

Figure 6. *Cont.*

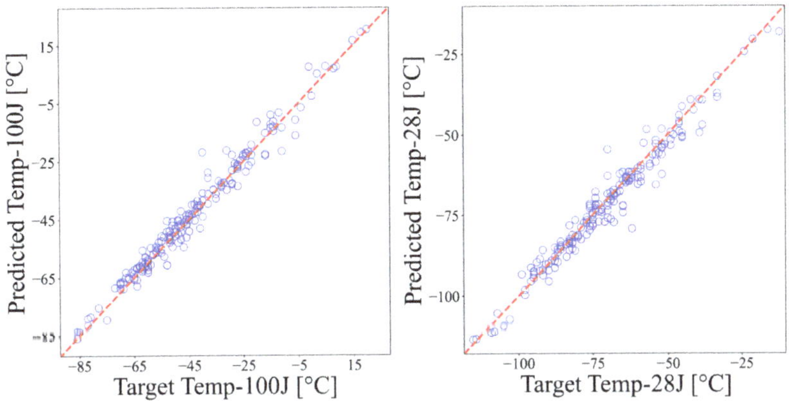

(e) (f)

Figure 6. Performance of the prediction model: (**a**) yield strength; (**b**) ultimate tensile strength; (**c**) elongation; (**d**) reduction of area; (**e**) transition temperature at 100 J; (**f**) transition temperature at 28 J.

The accuracy of the prediction model was assessed using the coefficient of determination (R^2) defined in Equation (6). The coefficient of determination reflects how well the regression model explains the variance between the predicted and actual data, with values ranging from 0 to 1. A value closer to 0 indicates weak explanatory power, while a value closer to 1 indicates that the model accurately captures the actual data. The results of the numerical evaluation of the prediction model's performance using the coefficient of determination are presented in Table 3. The findings indicate that the ANN model, developed from experimental data, is suitable for actual predictions of YS, UTS, Temp-100J, and Temp-28J. However, predicting El and RA proved challenging due to their weak correlation with the chemical compositions.

$$R^2 = 1 - \frac{\sum_i (y_i - \hat{y}_i)^2}{\sum_i (y_i - \overline{y})^2} \tag{6}$$

Table 3. Performance evaluation of the prediction model.

Element	YS	UTS	El	RA	Temp-100J	Temp-28J
	0.9805	0.9778	0.8208	0.6685	0.9739	0.9620

4.4. Explainable AI with Shapley Values

To quantitatively interpret how the ANN model predicts mechanical properties and to evaluate the relative importance of each input variable (chemical composition), an explainable artificial intelligence (XAI) method based on Shapley values was applied. This approach, grounded in cooperative game theory, fairly assigns each input variable's contribution to the predicted output, and is known to be effective in enhancing model interpretability even in complex multivariate regression problems [42,43].

Shapley values mathematically define the importance of each variable by calculating its average marginal contribution to the model's output. This can be expressed as follows:

$$\varphi_i(f, x) = \sum_S \frac{|S|!(|N| - |N| - 1)!}{|N|!} [f(S \bigcup \{i\}) - f(S)] \tag{7}$$

Here, φ_i denotes the Shapley value of feature i, N represents the set of all features, S is any subset of features excluding i, and $f(S)$ refers to the model prediction based solely on the input subset S. This formula calculates the average contribution of feature i to the model prediction across all possible combinations in which it is newly added. It provides a fair and consistent explanation even in the presence of interactions among variables.

Figure 7 presents a heatmap visualizing the average absolute Shapley values of 16 chemical elements across six predicted mechanical properties, indicating the extent to which each element contributes to the predictions.

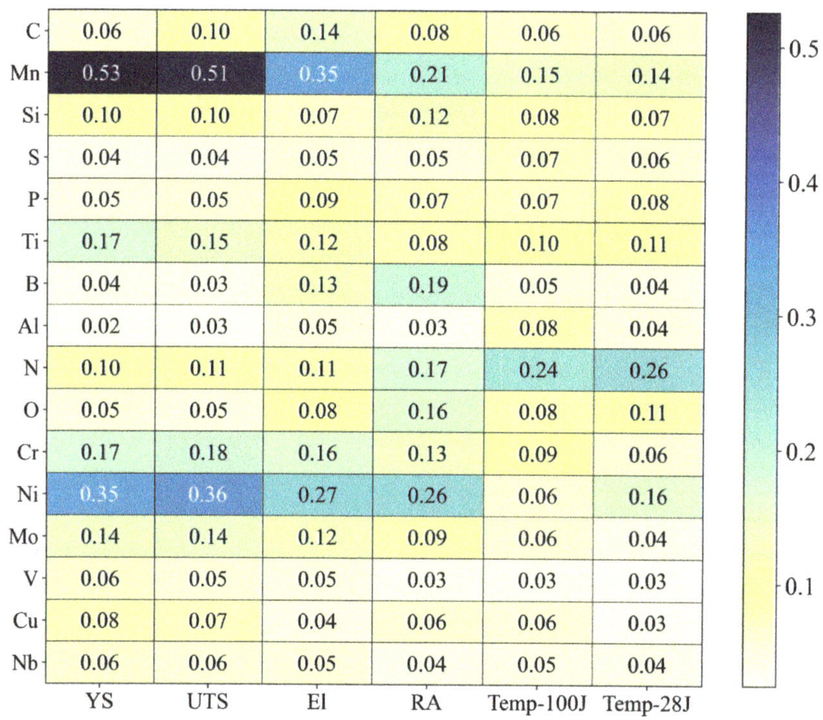

Figure 7. Shapley values representing the contribution of each chemical element to the prediction of mechanical properties.

The analysis revealed that Mn exhibited the highest contribution in predicting YS and UTS, confirming its critical role in determining the strength characteristics of weld metals. Ni also showed consistently high importance across all mechanical properties, with particularly notable contributions to the predictions of El, RA, and Temp-100J.

In contrast, N demonstrated high Shapley values for properties related to transition temperature, such as Temp-100J and Temp-28J, which are defined based on specific absorbed energies in Charpy V-notch impact tests. This suggests that N is an important factor in enhancing toughness. On the other hand, elements such as S, P, and Al generally exhibited low contributions, indicating that their influence on the model's predictions is relatively minor.

These results demonstrate that the ANN model does not function merely as a black box predictor but rather learns patterns consistent with metallurgical mechanisms. Furthermore, the findings provide valuable quantitative evidence for prioritizing alloying elements during the inverse design process, thereby improving design efficiency.

5. Result of Inverse Prediction Model

Once the ANN model describing the relationship between the chemical composition and mechanical properties of the weld is complete, it can be combined with a GA to explore combinations of chemical compositions that meet the target mechanical properties, as illustrated in Step 2 of Figure 2.

To apply the hybrid GA–ANN-based inverse prediction model, the target mechanical properties must first be determined. The case settings for the target mechanical properties used in this study are summarized in Table 4.

Table 4. Case settings for target mechanical properties.

Element	YS (N/mm^2)	UTS (N/mm^2)	El (%)	RA (%)	Temp-100J (°C)	Temp-28J (°C)
	489	571	30.2	78	−75	−102

The key parameters of the GA were set as follows. The population size was set to 100 to allow for a diverse range of composition combinations to be explored in each generation. The crossover probability was set to 0.8 to promote active recombination between parent chromosomes, while the mutation probability was set to 0.1 to prevent premature convergence to local optima and to ensure diversity in the search space. The algorithm was configured to run for a maximum of 100 generations, during which optimal composition combinations were iteratively derived. Table 5 summarizes the detailed GA parameters used in this study.

Table 5. Genetic algorithm parameters.

Parameters	Values
Population size	100
Crossover probability	0.8
Mutation probability	0.1
Maximum number of generations	100

After determining the mechanical properties, the next step is to load the pre-trained ANN model. This model follows the structure shown in Figure 4, where the chemical composition of the weld is entered into the input layer, and the corresponding mechanical properties are obtained from the output layer. The loaded ANN model acts as a black-box objective function within the inverse prediction model. It is important to note that since the trained ANN model acts as an objective function, it explores chemical composition combinations under the assumption of 100% prediction accuracy.

Each chromosome used in the GA consists of 16 types of weld chemical compositions. When a chromosome is input into the ANN, the mechanical properties for that particular combination are output. The relationship between the output properties and the pre-set target mechanical properties is evaluated using an L1 loss-based fitness function, as shown in Equation (8), where $\hat{y}_i$ represents the mechanical properties predicted by the inverse prediction model, and y_i represents the pre-set target mechanical properties.

$$L1\ loss = \sum_{i=1}^{n} |y_i - \hat{y}_i| \quad (8)$$

Figure 8 shows the evolution of the fitness value over the generations of the GA. The best and worst fitness values from 10 iterations of the inverse prediction are presented, along with the average value of these iterations, providing an overview of the trends observed during repeated trials.

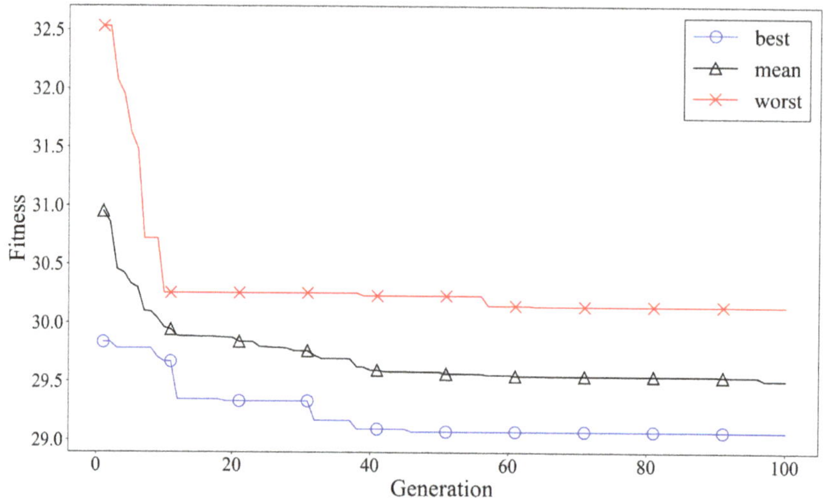

Figure 8. Fitness performance of the hybrid GA–ANN inverse prediction model over generations.

This analysis enables the identification of the chromosome with the smallest error between the predicted output and the target material properties, which is considered the optimized chemical composition to achieve the desired mechanical properties. The results of these iterations are summarized in Table 6.

Table 6. Optimized chemical compositions generated by the genetic algorithm.

Element	Trial 1	Trial 2	Trial 3	Trial 4	Trial 5	Trial 6	Trial 7	Trial 8	Trial 9	Trial 10	Unit
C	0.035	0.038	0.038	0.035	0.035	0.041	0.038	0.037	0.039	0.035	wt%
Mn	0.914	0.846	1.334	1.887	1.142	0.277	0.242	1.592	0.589	1.202	wt%
Si	0.731	1.081	0.823	0.437	0.926	0.036	0.016	0.648	0.019	0.033	wt%
S	0.0051	0.003	0.003	0.003	0.0073	0.003	0.003	0.008	0.003	0.0034	wt%
P	0.003	0.006	0.003	0.008	0.003	0.003	0.003	0.003	0.009	0.003	wt%
Ti	301.44	79.08	497.91	27.84	402.34	357.53	283.78	533.76	48.53	670.27	ppm
B	175.5	9.03	121.14	4.71	185.06	167.14	70.11	143.32	34.99	130.49	ppm
Al	256.44	56.83	558.31	5.13	262.96	35.41	25.38	411.21	5.27	155.95	ppm
N	120.38	111.47	136.52	152.45	152.41	195.66	44.95	227.48	126.85	249.63	ppm
O	283.86	235.56	253.61	273.91	254.92	304	260.96	220.8	423.59	244.35	ppm
Cr	0.061	0.065	0.245	0.262	0.091	0.104	0.062	0.036	0.104	0.266	wt%
Ni	2.732	0.078	4.020	0.254	0.048	0.058	0.058	2.743	1.933	0.061	wt%
Mo	0.055	0.121	0.062	0.522	0.041	0.024	0.030	0.017	0.040	0.005	wt%
V	119.15	25.26	24.23	530.65	191.64	307.96	144.5	27.04	356	122.74	ppm
Cu	0.13	0.02	0.7	0.12	0.22	0.08	0.02	0.34	0.06	0.08	wt%
Nb	7.32	539.03	12.17	7.15	19.69	3.05	630.31	62.81	5.72	43.33	ppm

Figure 9 visualizes the distribution of predicted mechanical properties by inputting the chemical compositions listed in Table 6 into the pre-trained ANN model. For each property, the target values are labeled, and the predicted property values over 10 trials are circled. The performance of the hybrid GA–ANN-based inverse prediction model was validated by assessing how well each chemical composition satisfied the target properties.

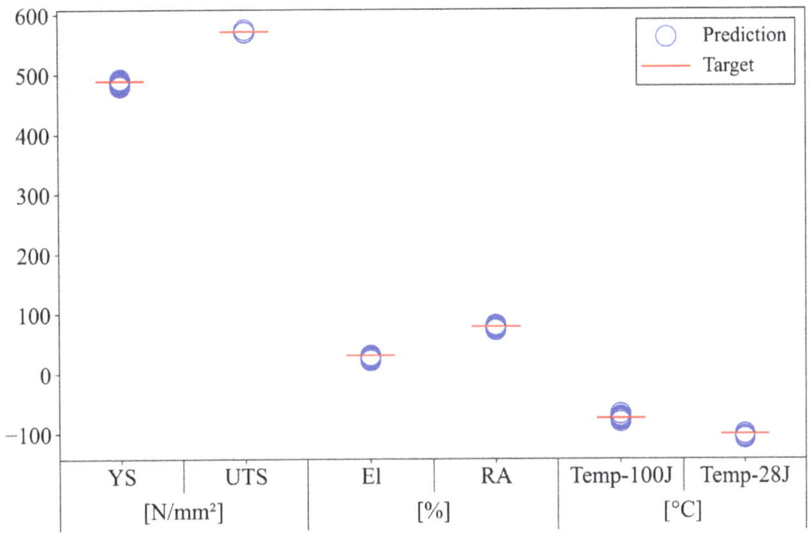

Figure 9. Mechanical properties corresponding to the suggested compositions.

6. Discussion

The hybrid GA–ANN framework proposed in this study has shown promise in effectively solving inverse design problems in welding alloy design. While traditional statistical analyses or individual ML models have proven effective in predicting properties for given compositions, they are limited in solving the inverse problem of deriving an optimal composition based on targeted material properties. This study overcomes these limitations by combining the predictive capabilities of ANNs with the optimization capabilities of GAs.

The ANN prediction model exhibited high accuracy, especially in key strength properties such as YS and UTS, while also maintaining a high level of prediction accuracy for Temp-100J and Temp-28J. This indicates that the influence of the chemical composition of SMAW welds on these properties can be effectively captured. Conversely, the prediction accuracy for El and RA was somewhat lower, as these properties are influenced by various factors, including welding process variables, cooling rates, and microstructure formation, and cannot be explained by chemical composition alone. This suggests the need to include additional physical variables or microstructure data for future model improvements.

To interpret the prediction results and ensure model transparency, this study applied an XAI approach based on Shapley values. This method enabled the quantitative evaluation of how much the ANN model relies on each chemical element when predicting mechanical properties. The analysis showed that Mn and Ni made the largest contributions to the predictions of YS and UTS, while N played a significant role in the predictions of Temp-100J and Temp-28J. These findings demonstrate that the ANN model can be extended beyond a traditional black-box predictor to an interpretable model through Shapley value-based analysis, where the contribution of each input variable can be quantitatively assessed. This enhances the model's reliability and transparency, reinforcing its potential as a decision-support tool in practical alloy composition design.

The GA-based inverse design process gradually converged to the optimal solution through repeated generation changes and mutation operations, with consistent results observed over 10 iterations. During this process, the L1 loss-based fitness function effectively minimized the absolute error between the target and predicted values, while the mutation operation helped prevent the algorithm from falling into a local minimum. Notably, several trials yielded compositional combinations that closely matched the target properties, demonstrating the reliability and practicality of the proposed framework.

The GA–ANN-based approach proposed in this study can serve as an effective tool for exploring global optima in composition design problems. However, alternative optimization algorithms such as Particle Swarm Optimization (PSO) and Bayesian inverse design methods may offer relative advantages in terms of convergence speed and quantification of predictive uncertainty. In future work, we plan to develop this framework further by comparing the performance of various optimization techniques depending on the problem characteristics, with the aim of identifying the most suitable algorithmic combinations. In particular, comparative studies with advanced evolutionary algorithms such as L-SHADE and EBLSHADE are also being considered as future research directions.

Furthermore, this study aimed not only to evaluate model performance but also to consider its potential application in real-world manufacturing environments. Defining target properties and identifying compositions that meet these requirements are closely related to the development of welding materials and prototype design. The approach of this study effectively reflects these practical needs. Such a model can serve as an intuitive and valuable design tool for both researchers and industry and is expected to significantly contribute to reducing the design cost and time for high-performance materials.

7. Conclusions

This study developed a hybrid AI-based model that can simultaneously predict the mechanical properties and perform inverse design of SMAW weld metals. First, based on a large-scale database provided by Dr. Glyn M. Evans, a multilayer ANN model was constructed, using the chemical composition of 16 elements as input and six mechanical properties as output. Bayesian optimization was used to fine-tune the model structure, resulting in good prediction accuracy.

Additionally, this study demonstrated the interpretability of the ANN model's predictions through an explainable AI analysis based on Shapley values. By quantitatively assessing the importance of each input element, the proposed approach contributes to enhancing the model's reliability and its practical applicability as a design tool.

Next, an inverse design framework was constructed by combining a GA with the trained ANN model. The GA iteratively refined the elemental composition to minimize the error between the target mechanical properties and the ANN predictions, thereby exploring various optimal combinations. In all 10 iterations of the inverse prediction, the compositional combinations closely matched the target properties, demonstrating the model's stability and effectiveness.

This hybrid GA–ANN approach represents a significant turning point beyond traditional composition-property modeling, enabling goal-oriented weldment design. Particularly in the early stages of weldment development, this model serves as a powerful tool to reduce experiment-based trial and error while also improving design efficiency and cost-effectiveness. Furthermore, the methodology presented in this study can be applied not only to SMAW but also to various welding methods, such as GMAW and GTAW, and dissimilar materials. It is expected to be widely applicable in various fields, including the development of new high-performance alloys and the design of materials for extreme environments.

Several considerations should be taken into account for future studies. First, in addition to chemical composition, process variables such as heat treatment conditions, cooling rate, and welding current should be included as inputs to increase the model's explanatory power. Second, efforts should be made to improve the accuracy of physical property predictions by integrating microstructure data and ensuring the physical validity of result interpretations. Finally, after deriving the optimal composition, it is necessary to conduct stepwise studies to examine the model's applicability in practice through specimen fabrication and experimental validation.

Author Contributions: Conceptualization, J.K. and J.-H.K.; methodology, Y.I.P. and J.-H.K.; validation, Y.I.P.; formal analysis, T.Y.; investigation, T.Y.; writing—original draft preparation, T.Y. and J.-H.K.; writing—review and editing, J.-H.K. and Y.I.P.; visualization, T.Y.; supervision, J.K.; project administration, J.K.; funding acquisition, J.K. All authors have read and agreed to the published version of the manuscript.

Funding: This study has been conducted with the support of the Korea Institute of Industrial Technology as "Development of a remote manufacturing system for high-risk, high-difficulty pipe production processes (KITECH EH-25-0004)" and This study has been conducted with the support of the Ministry of Trade, Industry and Energy as "Development of manufacturing technology using automated production system for wind turbine support structures (RS-2024-00512715)".

Institutional Review Board Statement: Not applicable.

Informed Consent Statement: Not applicable.

Data Availability Statement: The data presented in this study are available on request from the corresponding author.

Conflicts of Interest: The authors declare no conflicts of interest.

References

1. Evans, G.M.; Bailey, N. *Metallurgy of Basic Weld Metal*; Woodhead Publishing: Cambridge, UK, 1997.
2. Arulsamy, A.N.; S, G.; Murugesan, B.; Chelladurai, S.J.S.; Selvaraj, M.K.; Palanivel, V.; Balcha, G. Experimental investigation on microstructure and mechanical properties of friction welded dissimilar alloys. *Adv. Mater. Sci. Eng.* **2022**, *2022*, 5769115. [CrossRef]
3. Rajkumar, T.; Radhakrishnan, K.; Rajaganapathy, C.; Jani, S.P.; Salmaan, N.U. Experimental investigation of AA6063 welded joints using FSW. *Adv. Mater. Sci. Eng.* **2022**, *2022*, 4174210. [CrossRef]
4. Bin Reyaz, M.S.; Sinha, A.N. An experimental investigation on mechanical characteristics and wear behaviour of TIG welded dissimilar aluminum alloys. *J. Adhes. Sci. Technol.* **2024**, *38*, 1211–1246. [CrossRef]
5. Evans, G.M. *Database—Weld Metal Composition and Properties*; 2015. [CrossRef]
6. Sampath, K. Analysis of a high-strength steel SMAW database. *Weld. J.* **2021**, *100*, 410–420. [CrossRef]
7. Varadarajan, R.; Sampath, K. Application of machine learning to regression analysis of a large SMA weld metal database. *Weld. J.* **2023**, *102*, 31–52. [CrossRef]
8. Evans, G.M.; Quintana, M.; Babu, S. Comments on the numerical models for correlating weld metal composition to microstructure and properties. *Weld. J.* **2024**, *103*, 137–141. [CrossRef]
9. Sampath, K.; Varadarajan, R. High strength steel weld metal properties: Metallurgical criteria and computational tools. *Weld. World* **2023**, *67*, 2081–2105. [CrossRef]
10. Sampath, K. Metallurgical design rules for high-strength steel weld metals. *Weld. J.* **2022**, *101*, 123–143. [CrossRef]
11. Dhas, J.E.R.; Dhas, S.J.H. A review on optimization of welding process. *Procedia Eng.* **2012**, *38*, 544–554. [CrossRef]
12. Kim, I.S.; Lee, M.G.; Jeon, Y. Review on machine learning based welding quality improvement. *Int. J. Precis. Eng. Manuf.-Smart Technol.* **2023**, *1*, 219–226. [CrossRef]
13. Park, S.; Kim, C.; Kang, N. Artificial neural network-based modelling for yield strength prediction of austenitic stainless-steel welds. *Appl. Sci.* **2024**, *14*, 4224. [CrossRef]

14. Bera, T.; Das, S. Application of artificial neural networks in predicting output parameters of gas metal arc welding of dissimilar steels. *Indian Sci. Cruiser* **2021**, *35*, 26–30. [CrossRef]
15. Payares-Asprino, C. Prediction of mechanical properties as a function of welding variables in robotic gas metal arc welding of duplex stainless steels SAF 2205 welds through artificial neural networks. *Adv. Mater. Sci.* **2021**, *21*, 75–90. [CrossRef]
16. Jung, I.D.; Shin, D.S.; Kim, D.; Lee, J.; Lee, M.S.; Son, H.J.; Reddy, N.S.; Kim, M.; Moon, S.K.; Kim, K.T.; et al. Artificial intelligence for the prediction of tensile properties by using microstructural parameters in high strength steels. *Materialia* **2020**, *11*, 100699. [CrossRef]
17. Mezher, M.T.; Pereira, A.; Trzepieciński, T.; Acevedo, J. Artificial Neural Networks and Experimental Analysis of the Resistance Spot Welding Parameters Effect on the Welded Joint Quality of AISI 304. *Materials* **2024**, *17*, 2167. [CrossRef] [PubMed]
18. Mezher, M.T.; Pereira, A.; Shakir, R.A.; Trzepieciński, T. Application of Machine Learning and Neural Network Models Based on Experimental Evaluation of Dissimilar Resistance Spot-Welded Joints between Grade 2 Titanium Alloy and AISI 304 Stainless Steel. *Heliyon* **2024**, *10*, e40898. [CrossRef]
19. Victoria, A.H.; Maragatham, G. Automatic tuning of hyperparameters using Bayesian optimization. *Evol. Syst.* **2021**, *12*, 217–223. [CrossRef]
20. Wu, J.; Chen, X.Y.; Zhang, H.; Xiong, L.D.; Lei, H.; Deng, S.H. Hyperparameter Optimization for Machine Learning Models Based on Bayesian Optimization. *J. Electron. Sci. Technol.* **2019**, *17*, 26–40. [CrossRef]
21. Dubey, S.R.; Singh, S.K.; Chaudhuri, B.B. Activation Functions in Deep Learning: A Comprehensive Survey and Benchmark. *Neurocomputing* **2022**, *503*, 92–108. [CrossRef]
22. Shafiq, M.; Gu, Z. Deep Residual Learning for Image Recognition: A Survey. *Appl. Sci.* **2022**, *12*, 8972. [CrossRef]
23. Shen, Y.; Wang, J.; Navlakha, S. A Correspondence Between Normalization Strategies in Artificial and Biological Neural Networks. *Neural Comput.* **2021**, *33*, 3179–3203. [CrossRef] [PubMed]
24. Salehin, I.; Kang, D.-K. A Review on Dropout Regularization Approaches for Deep Neural Networks within the Scholarly Domain. *Electronics* **2023**, *12*, 3106. [CrossRef]
25. Abdulkadirov, R.; Lyakhov, P.; Nagornov, N. Survey of Optimization Algorithms in Modern Neural Networks. *Mathematics* **2023**, *11*, 2466. [CrossRef]
26. Noh, S.-H. Analysis of Gradient Vanishing of RNNs and Performance Comparison. *Information* **2021**, *12*, 442. [CrossRef]
27. Tian, Y.; Zhang, Y. A Comprehensive Survey on Regularization Strategies in Machine Learning. *Inf. Fusion* **2022**, *80*, 146–166. [CrossRef]
28. Jung, K.-H.; Lee, J.-H. Experimental study on predicting the corrosion behavior of carbon steel in various corrosive environments using artificial neural networks. *J. Adv. Mar. Eng. Technol.* **2023**, *47*, 317–324. [CrossRef]
29. Lee, J.-Y.; Kim, S.-H.; Jeong, H.-B.; Lee, K.; Cho, K.; Lee, Y.-K. Inverse Design of High-Strength Medium-Mn Steel Using a Machine Learning-Aided Genetic Algorithm Approach. *J. Mater. Res. Technol.* **2024**, *33*, 2672–2682. [CrossRef]
30. Bhat, N.; Barnard, A.S.; Birbilis, N. Inverse Design of Aluminium Alloys Using Genetic Algorithm: A Class-Based Workflow. *Metals* **2024**, *14*, 239. [CrossRef]
31. Schaufelberger, L.; Blaskovits, J.T.; Laplaza, R.; Jorner, K.; Corminboeuf, C. Inverse Design of Singlet-Fission Materials with Uncertainty-Controlled Genetic Optimization. *Angew. Chem. Int. Ed.* **2025**, *64*, e202415056. [CrossRef]
32. Dos Reis, F.; Karathanasopoulos, N. Inverse Metamaterial Design Combining Genetic Algorithms with Asymptotic Homogenization Schemes. *Int. J. Solids Struct.* **2022**, *250*, 111702. [CrossRef]
33. Hu, Z.; Ding, J.; Ding, S.; Ma, W.W.S.; Chua, J.W.; Li, X.; Zhai, W.; Song, X. Machine Learning–Enabled Inverse Design of Shell-Based Lattice Metamaterials with Optimal Sound and Energy Absorption. *Virtual Phys. Prototyp.* **2024**, *19*, e2412198. [CrossRef]
34. Jin, L.; Yu, S.; Cheng, J.; Liu, Z.; Zhang, K.; Zhou, S.; He, X.; Xie, G.; Bodaghi, M.; Ge, Q.; et al. Machine Learning Powered Inverse Design of Strain Fields of Hierarchical Architectures. *Compos. Part B Eng.* **2025**, *299*, 112372. [CrossRef]
35. Chiu, Y.-H.; Liao, Y.-H.; Juang, J.-Y. Designing Bioinspired Composite Structures via Genetic Algorithm and Conditional Variational Autoencoder. *Polymers* **2023**, *15*, 281. [CrossRef]
36. Qin, L.; Huang, W.; Du, Y.; Zheng, L.; Jawed, M.K. Genetic Algorithm-Based Inverse Design of Elastic Gridshells. *Struct. Multidiscip. Optim.* **2020**, *62*, 2691–2707. [CrossRef]
37. Shirvani, A.; Nili-Ahmadabadi, M.; Ha, M.Y. A Deep Learning–Genetic Algorithm Approach for Aerodynamic Inverse Design via Optimization of Pressure Distribution. *Comput. Methods Appl. Mech. Eng.* **2024**, *429*, 117187. [CrossRef]
38. Liu, B.; Jin, W.; Lu, A.; Liu, K.; Wang, C.; Mi, G. Optimal Design for Dual Laser Beam Butt Welding Process Parameter Using Artificial Neural Networks and Genetic Algorithm for SUS316L Austenitic Stainless Steel. *Opt. Laser Technol.* **2020**, *125*, 106027. [CrossRef]
39. Wang, W.; Zeng, Q.; Li, C.; Li, M.; Cao, L.; Chen, G.; Cao, P. A Novel Inverse Analysis Method for Mechanical Parameter Acquisition in SiCf/SiC Composites and Its Application to Turbine Disc Damage Assessment. *Materials* **2025**, *18*, 160. [CrossRef]

40. Lapre, C.; Meng, F.; Hary, M.; Morin, P.; Tani, F.; Travers, J.C. Genetic Algorithm Optimization of Broadband Operation in a Noise-Like Pulse Fiber Laser. *Sci. Rep.* **2023**, *13*, 1865. [CrossRef]
41. Sonta, A.; Dougherty, T.; Jain, R. Data-Driven Optimization of Building Layouts for Energy Efficiency. *Energy Build.* **2021**, *238*, 110815. [CrossRef]
42. Lundberg, S.; Lee, S.-I. A Unified Approach to Interpreting Model Predictions. *arXiv* **2017**, arXiv:1705.07874. [CrossRef]
43. Lundberg, S.M.; Erion, G.G.; Lee, S.-I. Consistent Individualized Feature Attribution for Tree Ensembles. *arXiv* **2018**, arXiv:1802.03888. [CrossRef]

materials

MDPI

Article

Exploring FAST Technique for Diffusion Bonding of Tungsten to EUROFERE97 in DEMO First Wall

María Sánchez [1,2,*], Javier de Prado [1,*], Ignacio Izaguirre [1], Andrei Galatanu [3] and Alejandro Ureña [1,2]

1 Materials Science and Engineering Area, Escuela Superior de Ciencias Experimentales y Tecnología, Universidad Rey Juan Carlos, C/Tulipán s/n, 28933 Móstoles, Spain; ignacio.izaguirre@urjc.es (I.I.); alejandro.urena@urjc.es (A.U.)

2 Instituto de Investigación de Tecnologías Para la Sostenibilidad, Universidad Rey Juan Carlos, C/Tulipán s/n, 28933 Móstoles, Spain

3 National Institute of Materials Physics, Atomistilor 405 A, 077125 Magurele, Romania; gala@infim.ro

* Correspondence: maria.sanchez@urjc.es (M.S.); javier.deprado@urjc.es (J.d.P.)

Abstract: The European Fusion Reactor (DEMO, Demonstration Power Plant) relies significantly on joining technologies in its design. Current research within the EUROfusion framework focuses on developing materials for the first wall and divertor applications, emphasizing the need for suitable joining processes, particularly for tungsten. The electric field-assisted sintering technique (FAST) emerges as a promising alternative due to its high current density, enabling rapid heating and cooling rates for fast sintering or joining. In this study, FAST was employed to join tungsten and EUROFERE97 steel, the chosen materials for the first wall, using 50-µm-thick Cu foils as interlayers. Three distinct joining conditions were tested at 980 °C for 2, 5, and 9 min at 41.97 MPa to optimize joint properties and assess FAST parameters influence. Hardness measurements revealed values around 450 $HV_{0.1}$ for tungsten, 100 $HV_{0.1}$ for copper, and 390 $HV_{0.1}$ for EUROFER97 under all joining conditions. Increasing bonding time improved joint continuity along the EUROFER97/Cu and W/Cu interfaces. Notably, the 5 min bonding time resulted in the highest shear strength, while the 9 min sample exhibited reduced strength, possibly due to Kirkendall porosity accumulation at the EUROFER97/Cu interface. This porosity facilitated crack initiation and propagation, diminishing interfacial adhesion properties.

Keywords: tungsten; EUROFERE97; copper interlayer; diffusion bonding

1. Introduction

The imminent deployment of future fusion power plants brings to the forefront the critical challenges posed by their high-power density and the extreme conditions prevailing within the vessel, characterized by intense irradiation, physical and chemical erosion, and sputtering. Addressing these challenges necessitates the development of advanced materials. Tungsten emerges as the primary candidate for providing armor to the first wall and divertor components, requiring effective integration with structural steel (EUROFER97) to meet stringent mechanical and metallurgical specifications. Therefore, the joining technology between both materials needs to be investigated. Tungsten, as a refractory metal with the highest melting point among all metallic materials, presents inherent physical and chemical properties for this task (high melting point, sputtering resistance, and thermal conductivity [1–3]). However, the high corrosion rate under air atmosphere at high temperatures has limited its application in the industry. As a result, its joinability has not been widely explored. The same properties that make this material adequate for this task also limit the available joining technologies, and an important effort should be made by the scientific community to address this issue.

Different techniques have been studied to join W to EUROFER97, including diffusion bonding (DB) processes [4,5], laser welding [6], and brazing [7,8]. However, the large differences in the thermal expansion coefficients (CTEs) make the bonding between W and EUROFER97 a big challenge [9,10]. Conventional methods have different limitations [11–13]. As an illustration, traditional diffusion bonding necessitates high temperatures and extended bonding times, making it a time- and energy-intensive process. Similarly, brazing involves the use of a filler metal with a lower melting point than the base materials, potentially compromising the mechanical properties of the joint under elevated temperatures and falling short of meeting application requirements.

In contrast, the electric field-assisted sintering technique (FAST), also known as spark plasma sintering (SPS), presents a novel approach for the swift consolidation of powders. This technique employs a robust, pulsed DC flow through powders or dies to generate joule heat, offering distinct advantages such as rapid heating rates, shortened sintering times, and low energy consumption [14–16]. The fusion of diffusion bonding with FAST technology emerges as a promising strategy for bonding tungsten with EUROFER97. Remarkably, the electrical current utilized in the FAST process promotes the migration of metallic and refractory ions at the joining interface, positively influencing the interfacial bonding strength of the joints [17].

Various researchers have explored the application of the FAST technique in joining materials, similar to those investigated in this study. For instance, D. G. Liu et al. [18] investigated the joint between an oxide dispersion-strengthened tungsten-based material and TZM, achieving optimal conditions at 1500 °C and obtaining a remarkable tensile strength of 485 MPa. In another study, Galatanu et al. [19] delved into a multi-layered composite between copper and tungsten, synthesized through a FAST-joining route. Notably, in heterogeneous joints, the Joule effect causes the material with the higher electrical resistivity to be heated more than the better electrical conductor. In the case of W–Cu laminates, tungsten is more heated than copper, and heat flows from tungsten towards copper at the interface. Structural, thermophysical properties, as well as tensile and Charpy impact tests conducted in this investigation, underscored that W–Cu laminates produced via FAST processing are comparable to those obtained through diffusion bonding. M. Naderi et al. [20] conducted investigations on the interaction between steel and copper, characterizing the microstructures and mechanical properties of Cu-AISI4140 steel solid-state joints created through spark plasma welding. Their examination included a comparative analysis of the process with and without a mold, revealing a notable reduction in unjoined areas and the formation of micropores confined at the joining interface. This modification contributed to an enhancement in the joining strength, escalating from 42 MPa to 90 MPa.

The FAST technique has also been used to explore functionally gradient material (FGM) fabrication. A. Pasha et al. fabricated a multilayer material based on Cr, Ti, Fe, Ni, Co, and Cu elements via Spark Plasma Sintering. The proportion of elements of each layer was gradually modified to mitigate the stress generated during the joining process of the components [21,22]. Other works explored the formation of grade interlayers constituted by seven layers of W/Cu functionally graded material (100 W, 80W-20Cu, 60W-40Cu, 50W-50Cu, 40W-60Cu, 20W-80Cu, 100Cu, by wt.%) by a spark plasma sintering process (SPS). The results indicated the consecution of fine microstructure within each layer with good interface bonding [23,24].

The current study introduces varied conditions for diffusion bonding aimed at joining tungsten to EUROFER97 for the first wall of DEMO. The diffusion-bonded joints underwent comprehensive characterization, employing scanning electron microscopy for microstructural analysis and microhardness, as well as shear tests for mechanical assessment.

2. Materials and Methods

2.1. Materials

Pure tungsten and EUROFER97 were the base materials. W was supplied by Plansee (>99.97 wt.% purity, Breitenwang, Austria). EUROFER97 had a standard composition and microstructure with the following nominal chemical composition (wt.%): 0.11 C, 8.7 Cr, 1.0 W, 0.1 Ta, 0.19 V, 0.44 Mn and 0.004 S [25]. The pure Cu interlayer was supplied by

Lucas-Milhaupt (with the commercial name *CDA 102K*, Cudahy, WI, USA, EEUU) and had a thickness of 50 μm. The interlayer had a composition of copper 99.95% Min, Oxygen 0.0010% Max and Other Elements (Total) 0.05% Max.

2.2. Joining by Field-Assisted Sintering Technique

The joining of W-EUROFER97 specimens was conducted using an FCT SPS D50 (Spark Plasma Sintering) equipment produced by *FCT Systeme GmbH* from Rauenstein, Germany. As the equipment allows only for integer values of force, an approach to achieve finer pressure tuning involved joining four samples simultaneously for each condition. The samples comprised blocks of W (6 × 6 × 5) mm, Cu foil (6 × 6 × 0.05) mm, and blocks of EUROFERE97 (6 × 6 × 5) mm. Prior to assembly, all pieces underwent thorough cleaning with 99.9% ethanol.

To facilitate alignment in the molds, a specialized tool was created and positioned between the pistons. An initial force of 2 kN was applied at room temperature to secure the pieces together. Subsequently, the tool was removed, and the die was raised to close the mold. The joining process unfolded within a low Ar atmosphere (<1 Pa), achieved through three vacuum/Ar-purging sequences.

The direct current pulses applied to heat the samples had a duration of 3 ms, with a 3 ms pause between pulses. The joining cycle encompassed a cold pressing step, reaching approximately 27 MPa, followed by a controlled heating phase, at a constant pressure, up to 450 °C, progressing at a rate of 100 °C/min. Simultaneously, the temperature increased at a rate of 40 °C/min up to 900 °C, accompanied by an elevation of pressure to 41.67 MPa. The final heating stage at this pressure reached up to 980 °C. Three dwell times were implemented: 2, 5, and 9 min. The selected bonding temperature ensures the austenitization of the steel during the bonding process (A_{1c} = 890 °C [25]), which will allow us to recover the hardness of the steel by applying a subsequent post-bonding heat treatment. The selected bonding pressure was set from 27 to 45 MPa, according to the literature review, where some works indicated optimal initial pressures around 25 MPa [19]. On the other side, as shown by Naderi et al. in Ref [20], for Cu steel joints, a pressure of 40 MPa achives enough plastic deformation, at this bonding temperature, to enhance the metallic contact. However, our FAST equipment allows only for force variation in 1 kN steps from 2 kN to 50 kN. The closest value available was 41.67 MPa, taking into account the joined surface area.

The cooling process was executed at a rate of 50 °C/min. Initially, there was a stepwise descent to 700 °C, maintaining constant pressure, followed by a gradual reduction in both temperature and pressure. Figure 1 shows a scheme of the joining process conditions for the 9 min sample.

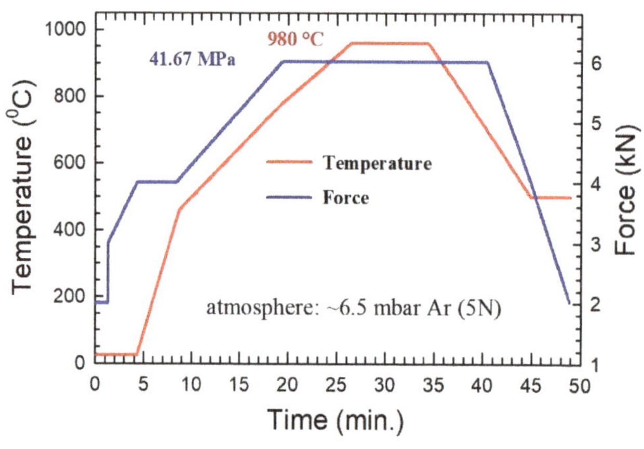

Figure 1. Scheme of the joining process conditions for the 9 min sample.

2.3. Characterization Techniques

The microstructural analysis of diffusion-bonded sample cross-sections was conducted using scanning electron microscopy (SEM, *S3400 Hitachi*, Tokyo, Japan) equipped with energy dispersive X-ray analysis (EDX). Metallographic preparation of the samples followed the standard polishing technique. Mechanical properties of the joints were assessed through Vickers microhardness and shear tests. Microhardness profiles across W-EUROFER97 joints were generated using an *MHV-2 SHIMADZU* (Tokyo, Japan) apparatus. A 100 g load was applied for 15 s, with three measurements obtained at each position. Distances between neighboring indentations exceeded three times the residual imprint sizes.

Shear strength values were determined using a shear fixture positioned between compression platens in a universal testing machine (*Zwick Z100*, Ulm, Germany) at a speed of 1 mm/min. The fixture was specifically designed to align the shear force with the bonding interface, maintaining the sample position so pure shear forces could be applied. To ensure accuracy, three samples from each condition underwent measurements.

3. Results

3.1. Microstructural Analysis

The microstructures of the three bonded joints under diffusion bonding (DB) conditions (980 °C for 2, 5, and 9 min at 41.97 MPa) are illustrated in Figure 2. Joints bonded with the shortest duration show the consecution of a continuous interface along the W/Cu region, with only isolated bonding zones evident at the EUROFER97/Cu interface (Figure 2a). These interface discontinuities suggested that the mechanisms of creep and interfacial diffusion were not conducive at the EUROFER97/Cu interface. Microstructural analysis revealed the presence of cracks along the interface, likely formed during the cooling stage. The dissimilarity in thermal expansion coefficients between Fe and Cu, combined with the insufficient interfacial strength, may contribute to this phenomenon. This fact is commonly observed when materials with a different coefficient of thermal expansion are joined and subjected to thermal processes. The different volume expansion of both materials generates internal stresses, which could lead to a crack nucleation or complete fracture of the joint/component in cases where the presence of low toughness phases or materials could not relieve the residual stress [21,22,26]. In this specific case, the lower adhesion properties of the EUROFER97/Cu interface promote the nucleation and propagation of the cracks following this interface.

Figure 2b,d,e reveal continuous and smooth variations in the Fe and Cu profiles under all three bonding conditions, suggesting interdiffusion between the materials across the interface. This interdiffusion predominantly occurred along grain boundaries, extending over a few microns. Furthermore, copper diffuses into the bulk grain, a phenomenon that will be elaborated upon in the subsequent microstructural discussion. Particularly, the W/Cu interfaces undergo a similar process; however, in this instance, the dissolution of W, and its subsequent precipitation, plays a pivotal role in shaping the observed smooth interface.

EDS punctual analyses carried out on both sides of both interfaces at distances of 0.5 microns form the interface line, indicating compositions of 89Fe-8Cr-3Cu at.% and 87Cu-12Fe-1Cr at.% at the EUROFER97 and Cu filler sides for the EUROFER97–Cu interface, respectively. In the case of the W–Cu interface, compositions of 75Cu-25W at.% and 91W-9Cu at.% for the Cu and W side, respectively, were detected. This information confirms the interdiffusion process of the elements that constitute the joint.

The extension of bonding time, specifically up to 5 and 9 min, significantly promoted the bonding process, effectively preventing the occurrence of cracking at the EUROFER97/Cu interface, as illustrated in Figure 2c and Figure 2e, respectively. The analysis of the W/Cu interface demonstrated the successful attainment of complete metallic continuity under all three FAST conditions. Despite the inherent low-temperature immiscibility of W and Cu, coupled with the refractory nature of W, a distinct interface emerged from the microstructural perspective. Remarkably, no metallurgical interactions, such as interdiffusion or the formation of third phases, were observed. Nonetheless, the plastic deformation of the copper filler at the bonding temperature facilitated the crucial establishment of intimate contact, raising metallic continuity and promoting the formation of metallic bonds.

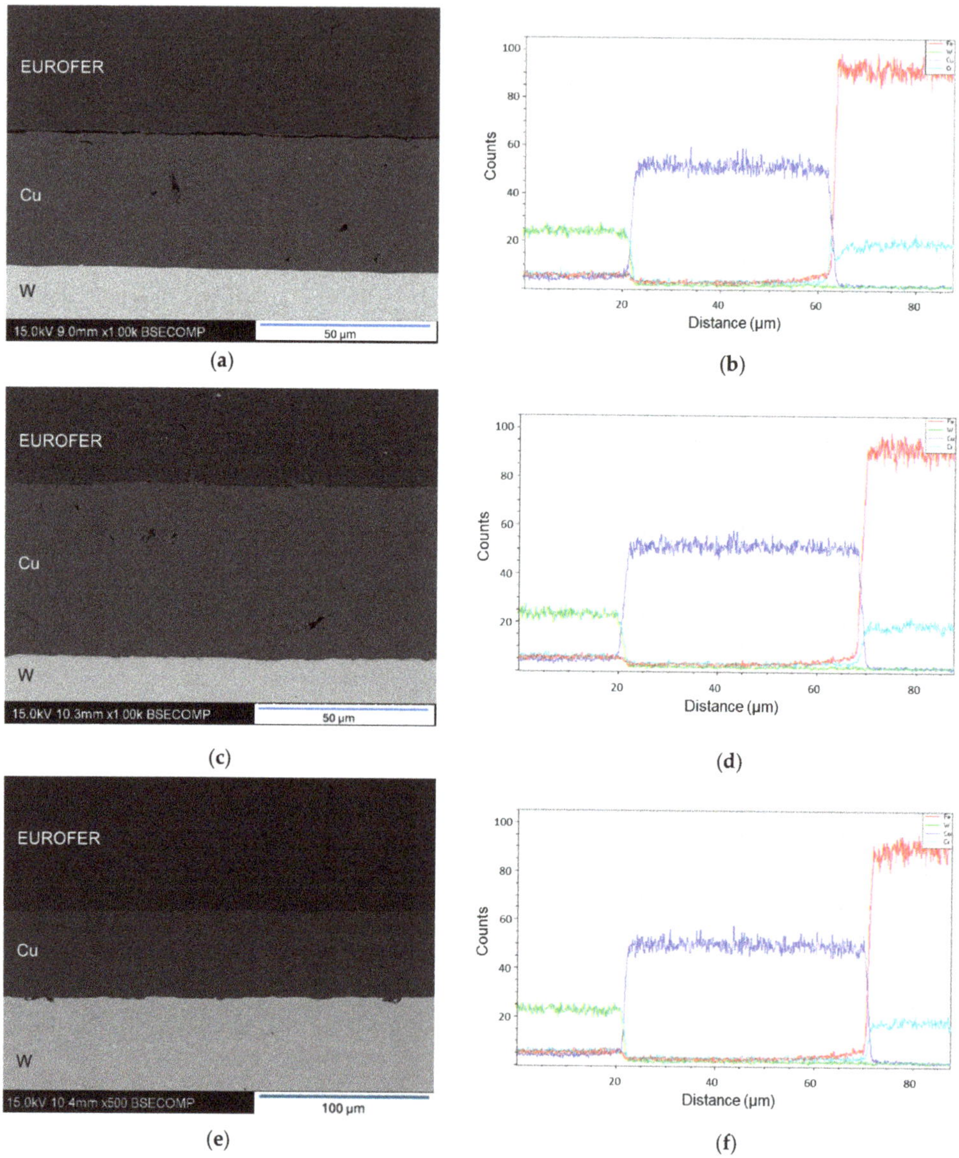

Figure 2. SEM micrographs of the W-EUROFER97 diffusion bonded joints at different conditions: (**a**) 2 min, (**c**) 5 min, and (**e**) 9 min. (**b**,**d**,**f**) Elemental diffusion distribution curves (Fe in red, Cr in blue, Cu in cyan and W in green) obtained from line scan analysis of EDX across the different joints corresponding to joining conditions (**a**), (**c**), and (**e**), respectively.

In addition, the lack of miscibility can be overcome by increasing the temperature up to temperatures close to the copper melting point because diffusion paths could be created.

J. Zhang et al. [27] successfully constructed a metallurgical bonding interface between W and Cu through diffusion, giving rise to W/Cu joints characterized by remarkable strength.

A deep examination of both interfaces at higher magnification was conducted to identify any potential welding defects or discontinuities. While these areas hold significance in conventional welding techniques, their importance is further emphasized in the FAST technique. In these zones, the initial contact area is low, leading to an elevated current flow and a substantial increase in temperature due to the Joule effect. The scrutiny was concentrated on the samples subjected to 5- and 9-min bonding times, as they exhibited superior weldability.

Figure 3a,b depict images corresponding to the EUROFER97/Cu interface for the 5 min and 9 min DB durations, respectively. In particular, microporosity is evident in both samples, with localized occurrences in the 5 min sample and a more abundant presence in the 9 min sample, extending along the interface. In certain regions of the lengthiest DB duration, this porosity becomes continuous. The observed location and geometry of the porosity suggest the formation of Kirkendall porosity during the DB process. This phenomenon arises when a significant disparity in the diffusion coefficients of the involved elements within the parent crystalline structure is established. According to the literature, at a temperature of 980 °C, the diffusion coefficient of copper in gamma iron is reported to be 7.63×10^{-13} cm^2/s [28], whereas the diffusion coefficient of iron in copper at the same temperature is notably higher at 1.26×10^{-9} cm^2/s [29]. Consequently, the accumulation of vacancies at the interface leads to porosity along the interface line, a process known as the Kirkendall effect.

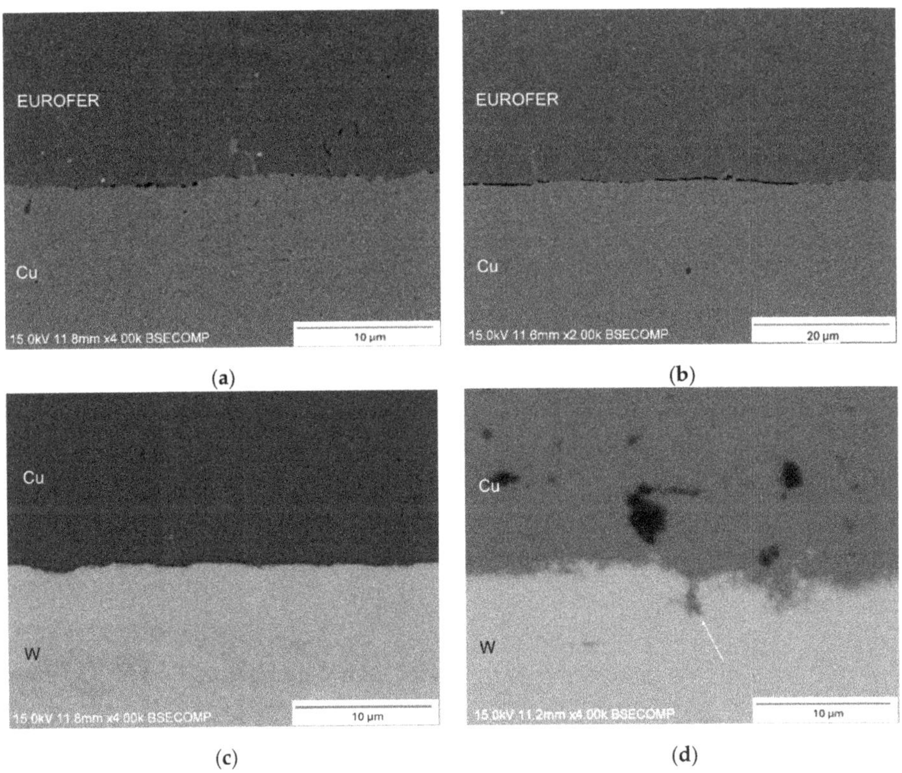

Figure 3. Detail of EUROFERE97/Cu interface of joints bonded at: (**a**) 5 min and (**b**) 9 min. Detail of W/Cu interface of joints bonded at: (**c**) 5 min and (**d**) 9 min.

The analysis of the W/Cu interface revealed local interaction between tungsten and copper, particularly in the 9 min sample (Figure 3d), where W diffuses and, during cooling, precipitates under these conditions within the Cu matrix. Conversely, a comparatively lower level of interaction was observed in the 5 min sample, attributed to its shorter exposure to the bonding temperature (Figure 3c). In the case of the extended DB duration, copper penetration along the W grain boundary was evident (indicated by the white arrow in Figure 3d). The inherent nature of FAST technology may elucidate this phenomenon. In the initial stages of the process, the contact area between W and copper is notably limited, causing the current to concentrate in these regions. This localized concentration of current could result in a temperature increase several hundred degrees higher than the bulk material, potentially leading to the local melting of the copper filler or an enhanced diffusion coefficient in the grain boundaries, facilitating copper migration. In addition, close to the melting point of copper, W could penetrate through a diffusion mechanism through copper grain boundaries, leading to partially solubilized W in the closest area of the joint interface. This W precipitates when the solubility decreases during the cooling stage of the process.

In general aspects, the obtained microstructure is similar to that obtained with other solid-state diffusion techniques. For example, E. Sal et al. studied the use of the HIP technique to join W to EURFOER97. They did not report extensive diffusion phenomena obtaining well-defined interfaces between the filler and the base materials. However, the intrinsic characteristic of this technique implies the application of longer joining times [30,31]. On the other hand, the use of liquid state joining techniques usually reports higher interactions at the bonding interfaces. For example, I. Izaguirre et al. reported copper penetration of the EUROFER97 grain boundary at the ERUFOER97/braze interface and the formation of a reaction layer at the W–braze interface, giving rise to a more complex microstructure [32,33].

3.2. Mechanical Characterization

Figure 4 illustrates the microhardness profiles extending from the tungsten base material to the EUROFER97 region across the joint. The profiles displayed remarkable similarities, indicating that the bonding time did not exert any discernible influence on the measured microhardness. The individual hardness values for tungsten, copper, and EUROFER97 were approximately 450, 100, and 390 $HV_{0.1}$, respectively. Tungsten hardness remained unaffected by the bonding process, maintaining its initial conditions. However, EUROFER97 exhibited an increased hardness compared to its as-received state (210 $HV_{0.1}$ [25]). The diffusion bonding temperature of 980 °C induced a hardness similar to that of EUROFER97 after undergoing austenization treatment and martensitic transformation when cooling rates higher than 5 °C/min is applied (~390 $HV_{0.1}$ [34]). In this work the utilized high cooling rate was effective in producing a non-equilibrium transformation of austenite into martensite, thereby elevating the material hardness. To restore EUROFER97 initial hardness, the joint necessitates a tempering treatment at 760 °C for 90 min, facilitating the recovery of the tempered martensite characteristic of this steel, restoring the dislocation density generated during the martensite formation and previous thermomechanical treatment, and producing the precipitation of the carbides that provide the precipitation hardening effect typical of this high temperature tempering process, allowing the application of the steel in high temperature conditions [35].

The analysis of the microhardness imprints carried out at the Cu filler did not reveal the presence of crack nucleation or propagation in the joint area (Figure 4b). Although this test is characterized by the application of forces in a reduce area and the nature of the Vickers imprints generated stress concentration points close to both interfaces (vortex of the imprint), both filler and interfaces have accommodated the stress without crack formation, which positively inform us about the toughness of the joint area and adhesion properties of the interfaces.

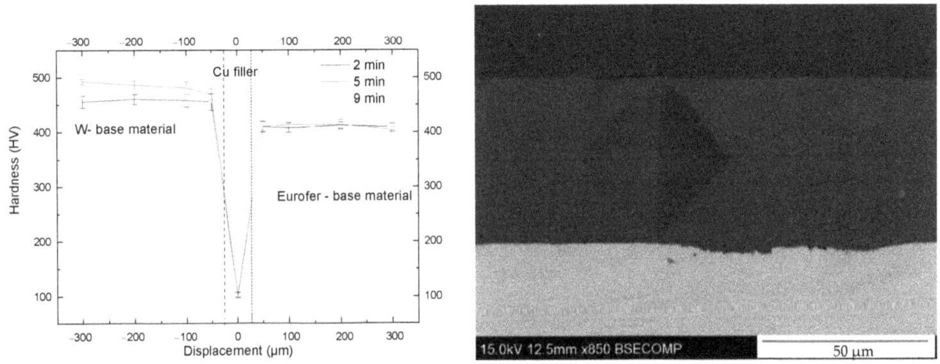

(**a**) (**b**)

Figure 4. (**a**) Vickers microhardness profiles of W-EUROFER97 joints under various evaluated conditions. (**b**) SEM image if the Vickers imprint carried out at the copper filler.

The shear strengths of the diffusion bonded joints were 48 ± 9 MPa, 120 ± 31 MPa, and 81 ± 14 MPa for samples bonded during 2, 5, and 9 min, respectively (Figure 5). Particularly, the joint bonded during the initial 2 min period exhibited lower strength, primarily attributed to the discontinuous EUROFER97/Cu interface. Mechanical properties showed a positive correlation with bonding time, as continuous interfaces were achieved for 5 min. However, the noticeable increase in Kirkendall porosity at the EUROFER97/Cu interface in the 9 min sample could contribute to a decrease in shear strength. As validated by microstructural examination, the accumulation of this porosity at the interface could result in the formation of a continuous discontinuity in certain areas. During the application of shear force, the formation and propagation of cracks along this layer are favored, leading to a reduction in the overall strength of the joint.

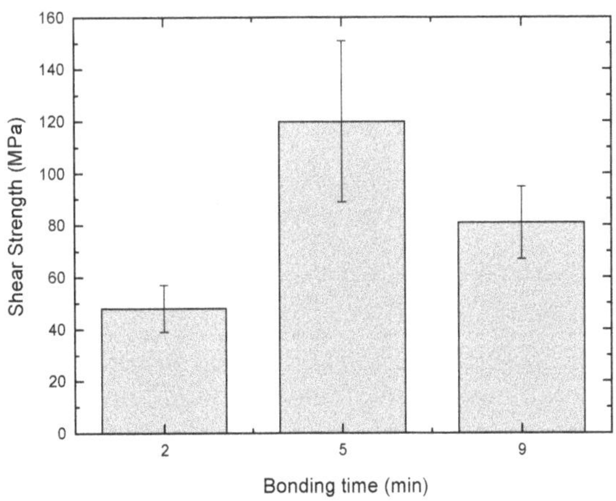

Figure 5. Shear strength of the W-EUROFER97 joints under various evaluated conditions.

As a consequence of the aforementioned mechanism, shear test failures primarily manifested along the EUROFER97/Cu interface, although it can shift to the W/Cu interface. For instance, Figure 6a,b illustrate the fracture surfaces of W and EUROFER97 components following shear strength tests. In particular, a substantial portion of the remaining interlayer was joined to the W base material, with the exception of the central region of the joint where the interlayer remained in the EUROFER97 base material. The presence of Kirkendall porosity at the EUROFER97/Cu interface elucidates the fracture propagation along this interface.

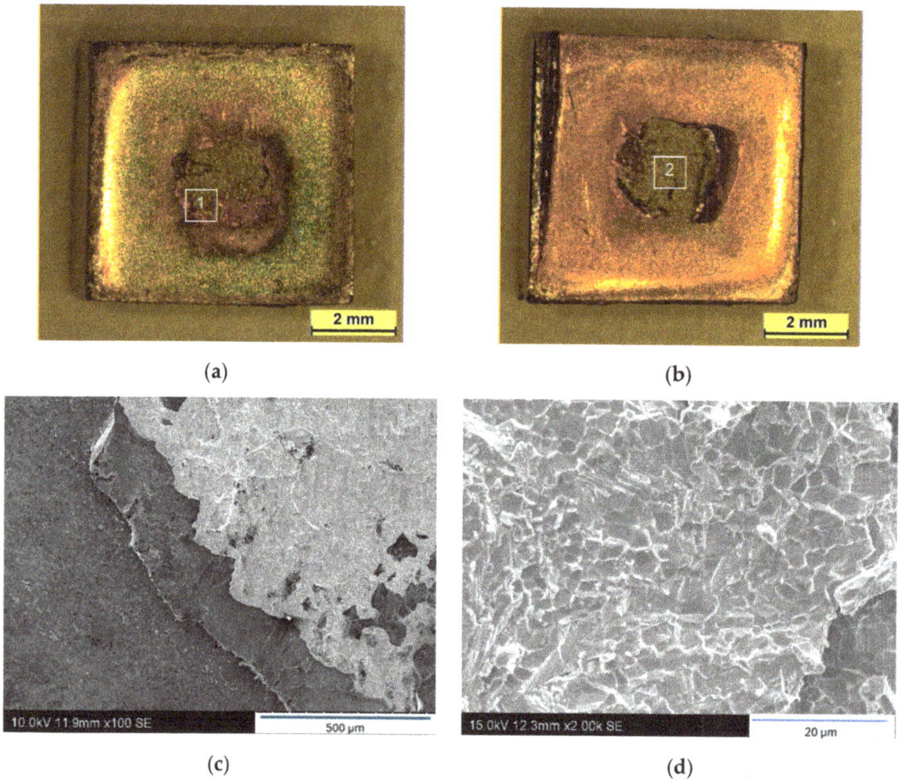

Figure 6. (**a**) EUROFER97 and (**b**) tungsten fracture surfaces of W-EUROFER97 joints using a Cu interlayer after the shear test. (**c**,**d**) detail the zones 1 and 2 marked in figures (**a**,**b**), respectively.

Utilizing scanning electron microscopy, the highlighted region in Figure 6a was examined, revealing the transition of failure from one interface to another. The contrasting material in the backscattering image of Figure 6c signifies W detachment from the base material during the fracture process, indicating the establishment of a strong interface between the interlayer and W base material. Consequently, when the fracture mechanism shifted from the EUROFER97/Cu to the W/Cu interface, the fracture did not propagate through the interface but rather through the W bulk material. Analysis of the W fracture surface unveiled an intergranular fracture pattern in the W substrate (Figure 6d).

The shear test results presented herein have been compared with those obtained through alternative joining methods utilizing the same Cu interlayer by our research group. Higher shear strengths were achieved through brazing and hot isostatic pressing (HIP) techniques [33,36,37]. Specifically, brazing at 1135 °C for 10 min yielded a shear strength of approximately 220 MPa. It is worth noting, however, that the molten Cu extensively infiltrated the grain boundaries of the EUROFER97 during brazing, leading to significant modifications in the microstructure of the steel near the brazed joint.

Conversely, the results obtained from HIP demonstrated joints with a superior shear strength of approximately 350 MPa, accompanied by minimal alteration in the microstructure of the steel. This outcome was achieved with an extended bonding time of 3 h.

After the shear test fracture, surfaces on both sides of the joint were analyzed by XRD to determine the crystalline structure and identify the phases that constitute the joint. Figure 7 shows the diffraction pattern, where the presence of Cu, Fe, and W has been identified on the ERUOFER97 side (Figure 7a) and only Cu on the W side (Figure 7b). This information confirms the presence of detached W from the base material, which remained adhered on the EUROFER97 side. The identification of Cu in the W side indicated that almost all copper remained on the W side, as previously indicated. This mechanical behavior has been previously observed in other works when high adhesion properties is achieved between the filler and W base material [38]. No further compounds, such as reaction products or intermetallic phases, have been identified.

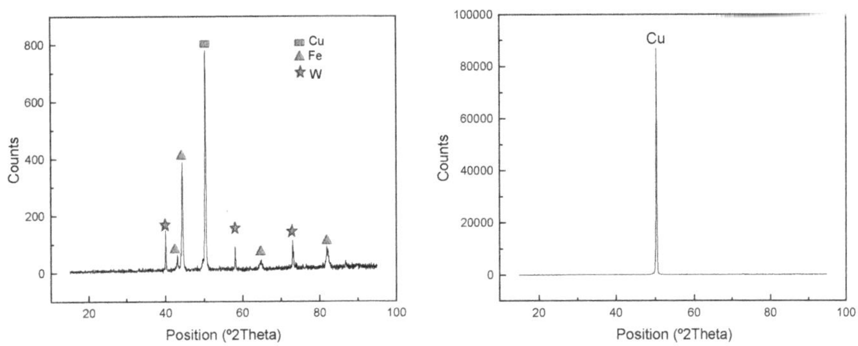

(a) (b)

Figure 7. X-ray diffraction patterns of (**a**) EURFOER97 and (**b**) W fracture surfaces.

4. Conclusions

Tungsten and EUROFER97 were effectively joined using a Cu interlayer material through the field-assisted sintering technique, employing bonding times exceeding 5 min. Shorter durations resulted in discontinuous joints, with only localized points joined at the EUROFERE97/Cu interface. The optimal condition was achieved with a 5 min dwell time at 980 °C. Prolonged dwell times led to the accumulation of Kirkendall porosity at the EUROFER97/Cu interface, giving rise to discontinuous regions in some areas. While the microstructure of the W base material remained unaltered, metallurgical phenomena, such as interdiffusion or dilution, occurred near both interfaces. These phenomena were intensified with longer dwell times at 980 °C.

Bonding time exhibited no discernible influence on the measured microhardness. The bonding process did not affect the hardness of tungsten. However, the EUROFER97 hardness exceeded its as-received condition due to the formation of untempered martensite, necessitating a tempering treatment for restoration.

The maximum shear strength was 120 MPa for the sample joined at 5 min. Higher bonding times increased the interaction at both interfaces, but more importantly, increased the interdiffusion mechanism at the EUROFER97/Cu interface, creating a preferential path for fracture propagation and potentially diminishing shear strength. While failure predominantly occurred at the weaker interface (EUROFER97/Cu), it eventually shifted to the W/Cu interface, causing fracture through the W base material due to the strength of this interface. Consequently, in these regions, detached W remained adhered to the EUROFER97 side of the fracture surface.

Author Contributions: Conceptualization, M.S. and A.U.; methodology, M.S., A.G., and J.d.P.; validation, A.U.; formal analysis, M.S. and J.d.P.; investigation, J.d.P., I.I., and A.G.; resources, M.S. and A.U.; data curation, M.S., J.d.P., and I.I.; writing—original draft preparation, M.S. and J.d.P.; writing—review and editing, A.U.; supervision, M.S. and A.U.; project administration, M.S.; funding acquisition, M.S. and A.U. All authors have read and agreed to the published version of the manuscript.

Funding: This work has been carried out within the framework of the EUROfusion Consortium, funded by the European Union via the Euratom Research and Training Programme (Grant Agreement No 633053 and Grant Agreement No 101052200—EUROfusion). Views and opinions expressed are, however, those of the author(s) only and do not necessarily reflect those of the European Union or the European Commission. Neither the European Union nor the European Commission can be held responsible for them.

Institutional Review Board Statement: Not applicable.

Informed Consent Statement: Not applicable.

Data Availability Statement: All data are contained within the article.

Conflicts of Interest: The authors declare no conflicts of interest.

References

1. Coenen, J.W.; Antusch, S.; Aumann, M.; Biel, W.; Du, J.; Engels, J.; Heuer, S.; Houben, A.; Hoeschen, T.; Jasper, B.; et al. Materials for DEMO and reactor applications—Boundary conditions and new concepts. *Phys. Scr.* **2015**, *2016*, 014002. [CrossRef]
2. Wurster, S.; Gludovatz, B.; Pippan, R. High temperature fracture experiments on tungsten–rhenium alloys. *Int. J. Refract. Met. Hard Mater.* **2010**, *28*, 692–697. [CrossRef]
3. Hohe, J.; Gumbsch, P. On the potential of tungsten–vanadium composites for high temperature application with wide-range thermal operation window. *J. Nucl. Mater.* **2010**, *400*, 218–231. [CrossRef]
4. Basuki, W.W.; Aktaa, J. Process optimization for diffusion bonding of tungsten with EUROFERE9797 using a vanadium inter-layer. *J. Nucl. Mater.* **2015**, *459*, 217–224. [CrossRef]
5. Zhong, Z.; Hinoki, T.; Nozawa, T.; Park, Y.H.; Kohyama, A. Microstructure and mechanical properties of diffusion bonded joints between tungsten and F82H steel using a titanium interlayer. *J. Alloys Compd.* **2010**, *489*, 545–551. [CrossRef]
6. Munez, C.J.; Garrido, M.A.; Rams, J.; Ureña, A. Experimental study of W-EUROFERE97 laser brazing for divertor application. *J. Nucl. Mater.* **2011**, *418*, 239–248. [CrossRef]
7. De Prado, J.; Sánchez, M.; Ureña, A. Evaluation of mechanically alloyed Cu-based powders as filler alloy for brazing tungsten to a reduced activation ferritic- martensitic steel. *J. Nucl. Mater.* **2017**, *490*, 188–196. [CrossRef]
8. De Prado, J.; Sánchez, M.; Ureña, A. Development of brazing process for W-EUROFERE97 joints using Cu-based fillers. *Phys. Scr.* **2016**, *T167*, 014022. [CrossRef]
9. Hirose, T.; Shiba, K.; Ando, M.; Enoeda, M.; Akiba, M. Joining technologies of reduced activation ferritic/martensitic steel for blanket fabrication. *Fusion Eng. Des.* **2006**, *81*, 645–651. [CrossRef]
10. Rakesh, P.; Davim, J.P. *Joining Processes for Dissimilar and Advanced Materials*; Woodhead Publishing: Soston, UK, 2021; ISBN 9780323853996.
11. Kazakov, N.F. *Diffusion Bonding of Materials*; Elsevier Science: Amsterdam, The Netherlands, 2013; pp. 143–154.
12. Tabernig, B.; Reheis, N. Joining of molybdenum and its application. *Int. J. Refract. Metals Hard Mater.* **2010**, *28*, 728–733. [CrossRef]
13. American Welding Society (AWS) C3 Committee on Brazing and Soldering. *Brazing Handbook*; American Welding Society: Miami, FL, USA, 1991.
14. Omori, M. Sintering, consolidation, reaction and crystal growth by the spark plasma system (SPS). *Mater. Sci. Eng. A* **2000**, *287*, 183–188. [CrossRef]
15. Munir, Z.A.; Quach, D.V.; Ohyanagi, M. Electric current activation of sintering: A review of the pulsed electric current sintering process. *J. Am. Ceram. Soc.* **2011**, *94*, 1–19. [CrossRef]
16. Guillon, O.; Gonzalez-Julian, J.; Dargatz, B.; Kessel, T.; Schierning, G.; Reathel, J.; Herrmann, M. Field-assisted sintering technology/Spark plasma sintering: Mechanisms, materials, and technology developments. *Adv. Eng. Mater.* **2014**, *16*, 830–849. [CrossRef]
17. Huang, Y.; Tillack, M.S.; Ghoniem, N.M. Tungsten monoblock concepts for the Fusion Nuclear Science Facility (FNSF) first wall and divertor. *Fusion Eng. Des.* **2018**, *135*, 346–355. [CrossRef]
18. Liu, D.G.; Ma, H.R.; Ruan, C.F.; Luo, L.M.; Zan, X.; Wang, Z.M.; Wu, Y.C. Effective joining between oxide dispersion strengthened tungsten-based material (ODS-W) and TZM alloy via spark plasma sintering technology. *Results Mater.* **2021**, *9*, 100175. [CrossRef]
19. Galatanu, A.; Galatanu, M.; Enculescu, M.; Reiser, J.; Sickinger, S. Thermophysical and mechanical properties of W-Cu laminates produced by FAST joining. *Fusion Eng. Des.* **2018**, *146*, 2371–2374. [CrossRef]

20. Naderi, M.; Toroghinejad, M.R.; Kermanpur, A. Spark plasma welding joining of copper- AISI4140 steel: Microstructures and mechanical properties. *Heliyon* **2023**, *9*, e21364. [CrossRef]
21. Pasha, A.; Rajaprakash, B.M. Functionally Graded Materials (FGM) Fabrication and Its Potential Challenges & Applications. *Mater. Today Proc.* **2022**, *52*, 413–418.
22. Fathi, R.; Wei, H.; Saleh, B.; Radhika, N.; Jiang, J.; Ma, A.; Ahmed, M.H.; Li, Q.; Ostrikov, K.K. Past and Present of Functionally Graded Coatings: Advancements and Future Challenges. *Appl. Mater. Today* **2022**, *26*, 101373. [CrossRef]
23. Chaubey, A.K.; Gupta, R.; Kumar, R.; Verma, B.; Kanpara, S.; Bathula, S.; Khirwadkar, S.S.; Dhar, A. Fabrication and characterization of W-Cu functionally graded material by spark plasma sintering process. *Fusion Eng. Des.* **2018**, *135 Pt A*, 24–30. [CrossRef]
24. You, J.-H.; Brendel, A.; Nawka, S.; Schubert, T.; Kieback, B. Thermal and mechanical properties of infiltrated W/CuCrZr composite materials for functionally graded heat sink application. *J. Nucl. Mater.* **2013**, *438*, 1–6. [CrossRef]
25. Fernandez, P.; Lancha, A.M.; Lapeña, J.; Hernandez-Mayoral, M. Metallurgical characterization of the reduced activation ferritic/martensitic steel EUROFERE97 97 on as received condition. *Fusion Eng. Des.* **2001**, *58*, 787–792. [CrossRef]
26. Papynov, E.K.; Shichalin, O.O.; Belov, A.A.; Buravlev, I.Y.; Portnyagin, A.S.; Azon, S.A.; Shlyk, D.K.; Buravleva, A.A.; Parot'kina, Y.A.; Nepomnyushchaya, V.A.; et al. Synthesis of Mineral-Like SrWO4 Ceramics with the Scheelite Structure and a Radioisotope Product Based on It. *Russ. J. Inorg. Chem.* **2021**, *66*, 1434–1446. [CrossRef]
27. Zhang, J.; Huang, Y.; Liu, Y.; Wang, Z. Direct diffusion bonding of inmiscible tungsten and copper at temperature close to Copper's melting point. *Mater. Des.* **2018**, *137*, 473–480.
28. Sen, S.K.; Dutt, M.B.; Barua, A.K. The diffusion of iron in copper and nickel in silver. *Phys. Stat. Sol. (a)* **1978**, *45*, 657–663. [CrossRef]
29. Salje, G.; Feller-Kniepmeier, M. The diffusion and solubility of copper in iron. *J. Appl. Phys.* **1977**, *48*, 1833–1839. [CrossRef]
30. Chen, X.; Bhattacharya, A.; Sokolov, M.A.; Clowers, L.N.; Yamamoto, Y.; Graening, T.; Linton, K.D.; Katoh, Y.; Rieth, M. Mechanical properties and microstructure characterization of EUROFERE9797 steel variants in EUROfusion program. *Fusion Eng. Des.* **2019**, *146 Pt B*, 2227–2232. [CrossRef]
31. de Prado, J.; Sal, E.; Sánchez, M.; García-Rosales, C.; Ureña, A. Microstructural and mechanical characterization of self-passivating W-EUROFERE97 joints processed by brazing technique. *Fusion Eng. Des.* **2021**, *169*, 112496. [CrossRef]
32. Izaguirre, I.; de Prado, J.; Sánchez, M.; Ureña, A. Wettability and microstructural evolution of copper filler in W and EUROFERE97 brazed joints. *Int. J. Adv. Manuf. Technol.* **2024**, *131*, 5905–5921. [CrossRef]
33. Sal, E.; de Prado, J.; Sánchez, M.; Ureña, A.; García-Rosales, C. Joining of self-passivating W-Cr-Y alloy to ferritic-martensitic steel by hot isostatic pressing. *Fusion Eng. Des.* **2021**, *170*, 112499. [CrossRef]
34. Chehtov, T.; Aktaa, J.; Kraft, O. Mechanical characterization and modeling of brazed EUROFERE97-tungsten-joints. *J. Nucl. Mater.* **2007**, *367–370 Pt B*, 1228–1232. [CrossRef]
35. Rieth, M.; Schirra, M.; Falkenstein, A.; Graf, P.; Heger, S.; Kempe, H.; Lindau, R.; Zimmermann, H. *EUROFERE 97. Tensile, Charpy, Creep and Structural Tests*; FZKA-6911; Forschungszentrum Karlsruhe GmbH Technik und Umwelt: Berlin, Germany, 2003.
36. De Prado, J.; Sánchez, M.; Ruiz, A.; Ureña, A. Effect of brazing temperature, filler thickness and post brazing heat treatment on the microstructure and mechanical properties of W-EUROFERE97 joints brazed with Cu interlayers. *J. Nucl. Mater.* **2020**, *533*, 152117. [CrossRef]
37. de Prado, J.; Sánchez, M.; Wirtz, M.; Pintsuk, G.; Du, J.; Linke, J.; Ureña, A. High heat flux performance of W-Eurofer brazed joints. *J. Nucl. Mater.* **2018**, *499*, 225–232. [CrossRef]
38. Tejado, E.; de Prado, J.; Tarancón, S.; Sánchez, M.; Martín, A.; Pastor, J.Y.; Ureña, A. Thermomechanical characterisation of W-Eurofer 97 brazed joints. *J. Nucl. Mater.* **2020**, *542*, 152504. [CrossRef]

materials

MDPI

Article

Interfacial Behavior During Reactions Between Sn and Electroplated Co–Zn Alloys

Chao-Hong Wang * and Che-Yang Lin

Department of Chemical Engineering, National Chung Cheng University, Chiayi 621301, Taiwan; jeff801104@gmail.com
* Correspondence: chmchw@ccu.edu.tw

Abstract: This study investigates the electroplating characteristics of Co-Zn alloy coatings with varying Zn contents (0.55 wt.%~8.8 wt.%) and their influence on intermetallic compound (IMC) formation during reactions with Sn solder. Co-Zn alloy coatings were successfully fabricated by electroplating using cobalt plating solutions with different concentrations of zinc sulfate. The results reveal anomalous co-deposition behavior, where the less noble Zn preferentially deposits over Co. Surface morphologies and microstructures evolve significantly with increasing Zn content, transitioning from columnar to dendritic structures. Zn incorporation into the Co lattice disrupts its crystallinity, leading to decreased crystallinity and partial amorphization. Liquid-state and solid-state interfacial reactions with Sn solder demonstrate that Zn content considerably influences IMC formation. In liquid-state reactions at 250 °C, lower Zn contents (0.55–4.8 wt.%) slightly enhance $CoSn_3$ growth. It exhibits a dense layered-structure without IMC spallation. In contrast, a higher Zn content (8.8 wt.%) significantly reduces IMC formation by approximately 50% and produces a duplex structure with two distinct layers. In solid-state reactions at 160 °C, the suppression effect becomes even more pronounced. The Co-0.55Zn deposit exhibits significant inhibition of $CoSn_3$ growth, while the Co-8.8Zn sample forms only a thin IMC layer, achieving a suppression rate exceeding 85%. These findings demonstrate that Zn doping effectively limits $CoSn_3$ formation during solid-state reactions and improves interfacial stability.

Keywords: interfacial reaction; cobalt; electroplating; co-deposition; lead-free solder; suppression

1. Introduction

Soldering technology plays a crucial role in microelectronic packaging, driving the advancement of various packaging technologies. Numerous tiny solder balls form solder joints that interconnect integrated circuit (IC) chips with IC substrates, such as in flip-chip packaging, or with printed circuit boards (PCBs) in ball grid array (BGA) technology [1–3]. As electronic devices continue to shrink in size, solder joints are being progressively miniaturized, raising increasing concerns about their reliability. To prevent the rapid dissolution of Cu pads by lead-free solders during reflow soldering, Ni is commonly used as a diffusion barrier. Among the various surface finishes developed for Cu pads, the electroless nickel immersion gold (ENIG) is one of the most widely adopted solutions [4–6]. Extensive studies have been conducted on the interfacial issues between lead-free solders and Ni or other related surface finish substrates [7–12].

In addition to Ni, Co has emerged as a promising alternative diffusion barrier due to its excellent electromigration resistance and the superior mechanical properties of Sn-Co

IMCs, which are beneficial for the reliability of solder joints [13–16]. Similar to electroless Ni(P), electroless Co(P) deposition has been developed for practical applications on solder pads [17–20]. The interfacial reactions between lead-free solders and Co or its related alloys have also been investigated [21–25], offering valuable insights for assessing their practical applicability. Nevertheless, Co exhibits faster IMC growth with solder compared to Ni [24,25]. Furthermore, in electroless Co(P) deposits, substantial IMC spalling has been observed during soldering, leading to rapid consumption of the Co(P) layer. This degradation can severely compromise the interfacial reliability of the solder joints [18,22,26]. As a result, ongoing research efforts are focused on addressing IMC spalling and suppressing IMC growth to improve the performance of Co-based diffusion barriers in soldering applications.

The minor addition of elements, such as Zn and Cu, into solder has been found to effectively suppress $CoSn_3$ growth at the solder/Co interface during soldering [16,27,28]. For instance, adding 0.5 wt.%Zn to Sn-based solder at 250 °C resulted in a significant reduction of approximately 75% in $CoSn_3$ growth [27]. This observation suggests that Zn doping in Co-Zn alloys may also affect the interfacial behavior of Sn-Co IMCs. This indicates that Zn may play a critical role in modifying interfacial IMC behavior. Inspired by this, Zn doping directly into the Co substrate may also influence IMC formation, although this has not been thoroughly investigated.

This study aims to investigate the interfacial reactions between Sn-based solder and Co-Zn alloy coatings with varying Zn contents. The Co-Zn coatings were fabricated via electroplating in a cobalt sulfamate-based bath containing different concentrations of zinc sulfate, with the Zn content controlled to be below 10 wt.%. Due to the known anomalous co-deposition behavior in Co-Zn systems [29–31], the less noble Zn may preferentially deposit over the more noble Co, resulting in a higher Zn content in the deposit than predicted. While the prior study has focused on the addition of Zn into solder, the effects of Zn incorporated directly into the substrate (i.e., the metallization layer) on interfacial IMC formation remain unclear. Therefore, a systematic investigation was conducted to examine how varying Zn concentrations in the Co-Zn coatings influence the growth behavior of intermetallic compounds. This work reveals that increasing Zn content in the Co-Zn coatings significantly suppresses $CoSn_3$ growth during soldering. A nucleation-based mechanism is proposed to explain the suppression behavior.

2. Materials and Methods

To fabricate Co-Zn alloy coatings onto Cu substrates, zinc sulfate ($ZnSO_4 \cdot 7H_2O$) was added to a basic Co electroplating bath based on a cobalt sulfamate ($Co(SO_3NH_2)_2$) system. The constituents of the Co electroplating solution are listed in Table 1. Four Co-Zn alloy coatings were prepared by introducing different amounts of $ZnSO_4 \cdot 7H_2O$, 0.1 g, 0.5 g, 1 g, and 2 g, corresponding to Zn^{2+} concentrations of 0.0017 M, 0.0087 M, 0.017 M, and 0.035 M, respectively, into 200 mL of the plating solution, as summarized in Table 2. Each plating solution was maintained in a 55 °C water bath and thoroughly stirred to ensure complete dissolution of $ZnSO_4 \cdot 7H_2O$. Before electroplating, a 0.5 mm-thick Cu substrate was cut into 10 mm × 15 mm pieces. Each specimen was mechanically polished with alumina powder and then ultrasonically cleaned in deionized (DI) water. The plating window was controlled to a size of 7 mm × 10 mm. During the Co-Zn electrodeposition process, the Cu substrate and a Co plate served as the cathode and anode, respectively, positioned 6 cm apart. Electroplating was carried out at 55 °C under a current density of 2 ASD (ampere per square decimeter) for 25 min. The theoretical deposited thickness was ~10 μm.

Table 1. Composition of the electroplating Co solution (per liter of solution).

Component	Concentration (g/L or mL/L)	Concentration (M)
Cobalt sulfamate solution ($Co(SO_3NH_2)_2$, Co > 180 g/L)	500 mL/L	1.527 M
Cobalt chloride ($CoCl_2$)	10 g/L	0.077 M
Boric acid (H_3BO_3)	40 g/L	0.647 M

Table 2. Formulation of the electroplated Co-Zn alloy coatings and results of EPMA compositional analysis.

Adding Amount of $ZnSO_4{\cdot}7H_2O$ (Based on 200 mL of Co Electroplating Solution)	$ZnSO_4{\cdot}7H_2O$ Concentration (M)	Zn Content in Co-Zn Coatings (wt.%)	Average Zn Content and Notation
0.1 g	0.00174 M	0.51–0.62	Co-0.55 wt.%Zn
0.5 g	0.0087 M	1.93–3.14	Co-2.5 wt.%Zn
1.0 g	0.0174 M	4.02–5.62	Co-4.8 wt.%Zn
2.0 g	0.035 M	8.49–9.08	Co-8.8 wt.%Zn

The top-view morphology of the Co-Zn deposits was examined using scanning electron microscopy (SEM), and their composition was analyzed with an electron probe microanalyzer (EPMA, JEOL JXA-8200, JEOL Ltd., Tokyo, Japan). The deposits were characterized by X-ray diffraction (XRD, Bruker D8) analysis with Cu-K_α radiation (λ = 0.154056 nm). The diffraction patterns were identified by comparison with reference patterns from the Joint Committee on Powder Diffraction Standards (JCPDS) database. For cross-sectional observation of the grain structure, the Co-Zn deposits were examined using focused ion beam (FIB) microscopy (FEI Quanta 3D FEG, Thermo Fisher Scientific, Waltham, MA, USA). The samples were lightly polished to obtain a flat surface. A protective Pt layer was then deposited, followed by FIB milling. The grain structure of the Co-Zn coatings was observed using ion-induced secondary electron (ISE) imaging.

For the soldering reaction, the Co-Zn deposits were polished to obtain a flat surface and then reacted with Sn at 250 °C for durations ranging from 10 s to 2 h. In addition to the soldering reaction, solid-state reactions were also investigated. To prepare the reaction couples, Sn solder was reflowed onto the Co-Zn deposits at 250 °C for 3 s to form an initial joint. Subsequently, the samples were aged at 160 °C on a hot plate for predetermined durations, such as 160 h. After the reaction, the samples were mounted in epoxy resin, and then ground and polished for metallographic analysis. The polished samples were lightly etched with an Sn etchant to enhance the visibility of the interfacial zone. The interfacial morphologies were then examined using back-scattered electron imaging (BEI) mode in SEM. The composition of the reaction phases was analyzed using EPMA. The thickness of the reaction phase was determined by calculating the total area of the phase using image analysis software (Optimas version 6.1, Optimas Corporation, Wood Dale, IL, USA) and dividing it by the measured interface length.

3. Results and Discussion

3.1. *Characterization of Electroplated Co-Zn Coatings*

As described in the experimental methods, four Co-Zn alloy coatings with varying Zn contents were successfully fabricated by electroplating, in which different amounts of zinc sulfate were added to 200 mL of a Co electroplating solution. The chemical compositions of the resulting coatings, determined by EPMA measurements at multiple locations across the cross-sections, are summarized in Table 2. The coatings were denoted as Co-0.55 wt.%Zn, Co-2.5 wt.%Zn, Co-4.8 wt.%Zn, and Co-8.8 wt.%Zn, respectively. As shown in Figure 1, the Zn content in the deposits increases nearly linearly with increasing Zn^{2+} concentration in the plating bath, corresponding to the amount of zinc sulfate added. The relatively narrow compositional ranges suggest that the Zn distribution within each coating is reasonably uniform, with no apparent segregation or compositional inhomogeneity observed.

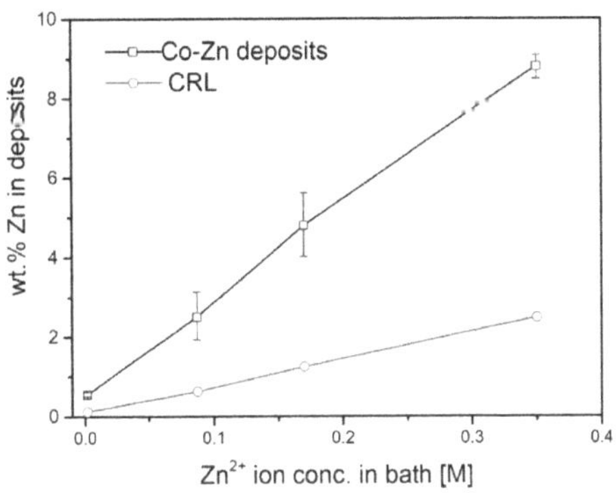

Figure 1. Zn concentrations in the Co-Zn deposits as a function of Zn^{2+} ion concentration in the electroplating bath.

In Figure 1, the compositional reference line (CRL) represents the predicted Zn content in the deposits, calculated based on the concentration ratio of Zn^{2+} ions to the total metal ion concentration (Co^{2+} and Zn^{2+}) in the plating solution. Notably, the experimentally measured Zn contents were consistently higher than the CRL, suggesting that Zn, being electrochemically less noble than Co, was preferentially deposited during the electroplating process. This behavior was characteristic of anomalous co-deposition [29,30]. The Zn contents in the deposits were approximately 4.4, 3.8, 3.8, and 3.5 times higher than the CRL values, respectively, showing a slight downward trend with increasing Zn^{2+} concentration in the plating bath.

The observed anomalous co-deposition behavior, where Zn exhibits a greater tendency to deposit than the more noble Co, can be attributed to several electrochemical factors, including localized pH increases and kinetic effects. Although Zn has a more negative standard reduction potential (−0.76 V for Zn^{2+}/Zn vs. −0.28 V for Co^{2+}/Co), it can still be preferentially deposited due to local pH changes near the cathode surface. The increase in local pH promotes the precipitation of $Zn(OH)_2$, which accumulates on the cathode surface and acts as a barrier, hindering the discharge of Co^{2+} ions and thus suppressing Co deposition [32–34]. Additionally, Zn exhibits faster deposition kinetics than Co, which further enhances its preferential deposition during the co-electroplating process [29].

Figure 2a–c shows the surface morphologies of electroplated Co-Zn deposits with Zn contents of 0.55 wt.%, 2.5 wt.%, and 8.8 wt.%, respectively. Significant changes in surface morphology were observed with increasing Zn content. A rough and irregular surface texture was observed in the Co-0.55Zn deposit, composed of large, angular, faceted grains approximately 1 μm in size with well-defined boundaries. In contrast, the Co-2.5Zn deposit displayed densely packed, elongated, needle-like grains, that formed a more interconnected and textured structure. Notably, two distinct grain size populations were observed: larger grains (~2 μm) and finer grains, with sizes of several hundred nanometers. As the Zn content increased, the grains became progressively finer, and the fraction of larger grains decreased. In the Co-8.8 wt.%Zn coating, the surface consisted of uniformly distributed fine grains, leading to a smoother and more homogeneous morphology.

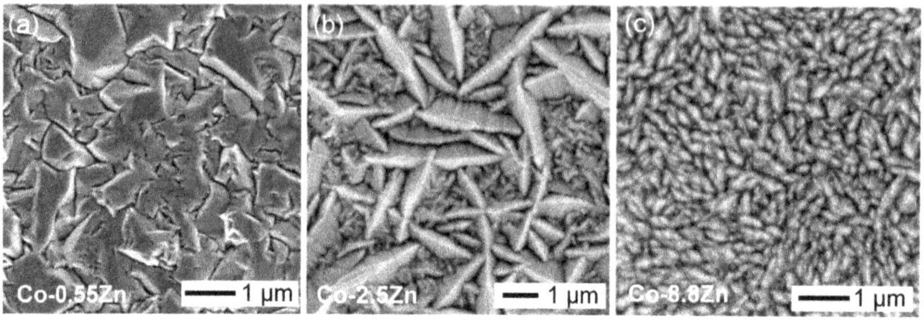

Figure 2. Top-view SEM images of the electroplated Co-Zn surfaces with different Zn contents: (**a**) Co-0.55Zn, (**b**) Co-2.5Zn, and (**c**) Co-8.8Zn.

Cross-sectional observations of the coatings were performed using FIB to investigate the structural evolution of Co-Zn coatings with varying Zn contents, as shown in Figure 3a–d. At low Zn content (0.55 wt.%), the coating exhibited coarse and straight columnar grains. As the Zn content increased to 2.5 wt.% and 4.8 wt.%, the columnar structure was maintained, but the grains became more disordered and refined. At the highest Zn content (8.8 wt.%), the columnar grains evolved into a finer dendritic morphology, suggesting that Zn incorporation promoted grain refinement and disrupted the typical columnar growth pattern. These microstructural changes were consistent with the variations observed in the surface morphologies as the Zn content increased.

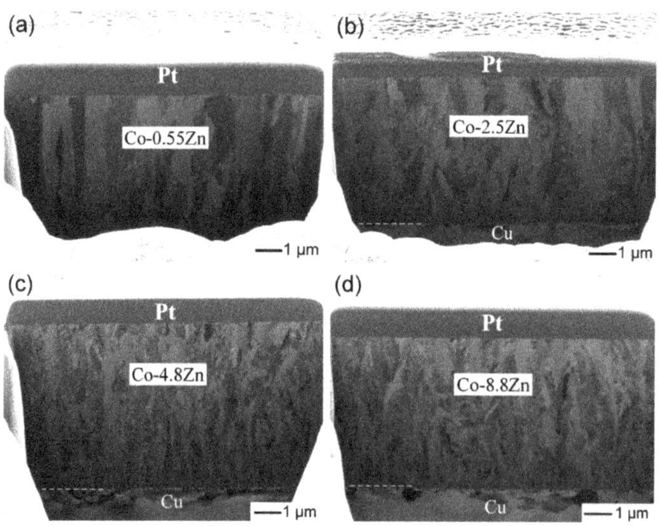

Figure 3. FIB cross-sectional images of the electroplated Co-Zn deposits: (**a**) Co-0.55Zn, (**b**) Co-2.5Zn, (**c**) Co-4.8Zn, and (**d**) Co-8.8Zn.

To further assess the crystallinity and crystallographic orientation of the Co-Zn deposits, XRD analyses were conducted, with a pure Co deposit used as a reference, as shown in Figure 4a–e. In the XRD pattern of the Co-0.55Zn deposit, the hcp-Co peaks (JCPDS #89-4308) were significantly weaker than those of the underlying fcc-Cu substrate (JCPDS #89-2838), indicating a substantial reduction in the crystallinity of the Co phase. Notably, a distinct broad hump was observed in the 2θ range of 10° to 25°, which was characteristic of an amorphous structure. This feature was absent in the pure Co deposit and was thus attributed to the incorporation of Zn atoms into the hcp-Co lattice, which disrupted long-range atomic ordering. Although both Zn and Co possess hcp crystal structures, the larger atomic radius of Zn induced severe lattice distortion, leading to a loss of crystallinity. Despite this reduction, a preferred orientation along the (110) plane of hcp-Co at a 2θ value of 75.8° remained observable, similar to that in the pure Co deposit.

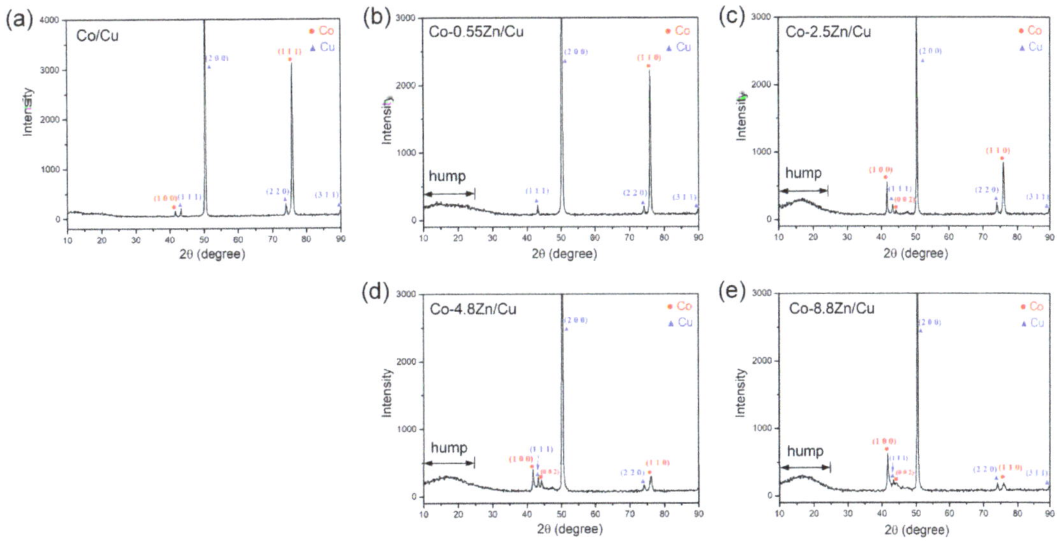

Figure 4. XRD patterns of the electroplated Co-Zn deposits on Cu substrate, (**a**) Co, (**b**) Co-0.55Zn, (**c**) Co-2.5Zn, (**d**) Co-4.8Zn, and (**e**) Co-8.8Zn.

As the Zn content increased to 2.5 wt.%, the intensity of the Co peaks further decreased, and the broad hump became more pronounced. This indicated that Zn incorporation into the Co lattice further reduced the crystallinity of the Co phase and induced partial amorphization. In addition to the (110) peak of hcp-Co, a prominent (100) peak appeared at 2θ = 41.6°. Similar features were observed in the XRD patterns of the Co-4.8Zn and Co-8.8Zn deposits. In contrast, in the XRD spectra, the (002) peak remained much weaker, suggesting that the Co-Zn coatings preferentially grew along the (100) and (110) planes, with the c-axis of the hcp-Co structure oriented parallel to the substrate. The XRD analysis revealed a noticeable reduction in crystallinity with increasing Zn content in the Co-Zn coatings. This structural change suggests the potential for enhanced hardness and strength, as amorphous or nanocrystalline structures typically hinder dislocation motion. However, it should be noted that such microstructural characteristics may also result in increased internal stresses and reduced toughness, which could negatively affect the mechanical reliability of the coatings.

3.2. Interfacial Reactions of Co-Zn Deposits with Sn at 250 °C

Figure 5a–f shows the interfacial microstructures of the Sn/Co-0.55Zn reactions at 250 °C for various durations, ranging from 10 s to 1.5 h. After just 10 s, a thin reaction layer was formed. Notably, the interface between the IMC and the Co-0.55Zn coating appeared irregular, suggesting slight dissolution of the Co-0.55Zn into the solder during the initial reaction stage. After 10 min of aging, a bright and significantly thicker IMC layer, approximately 15 μm, was observed. EPMA analysis indicated that the IMC consisted of

Sn-25.06 at.%Co and 0 at.%Zn, and was identified as the $CoSn_3$ phase, with no detectable Zn content. The IMC grew rapidly; by 30 min (Figure 5c), the layer reached ~31 μm in thickness. With further aging to 1 h (Figure 5d), the dense reaction phase thickened further to approximately 45 μm.

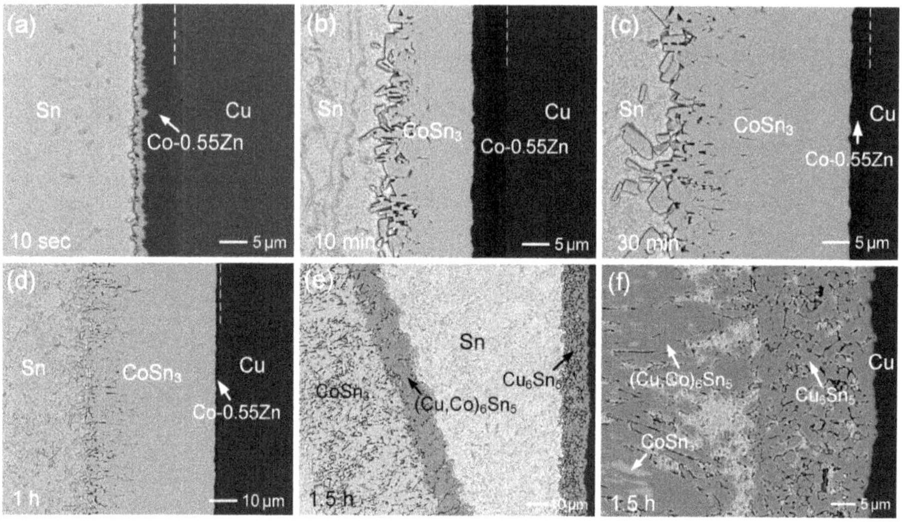

Figure 5. BEI micrographs of the Sn/Co-0.55Zn reaction at 250 °C for different durations: (**a**) 10 s, (**b**) 10 min, (**c**) 30 min, (**d**) 1 h, (**e**) 1.5 h, and (**f**) magnified view of the sample after 1.5 h.

For comparison, the reaction was also performed using electroplated pure Co, as shown in Figure 6a–c. As seen in Figure 6b, the $CoSn_3$ layer reached a thickness of approximately 34 μm. The IMC layers formed after 10 min and 30 min of aging were relatively thinner than those observed in the Co-0.55Zn reaction. These results suggested that the addition of 0.55 wt.%Zn enhanced the reaction kinetics and promoted IMC growth. In Figure 6a, the interface between the IMC and pure Co appeared relatively straight, further supporting the earlier suggestion that the Zn addition accelerated the dissolution of the Co-Zn alloy into molten Sn. After 2 h of aging (Figure 6c), the IMC layer became very thick (~55 μm), although a thin layer of the Co deposit still remained.

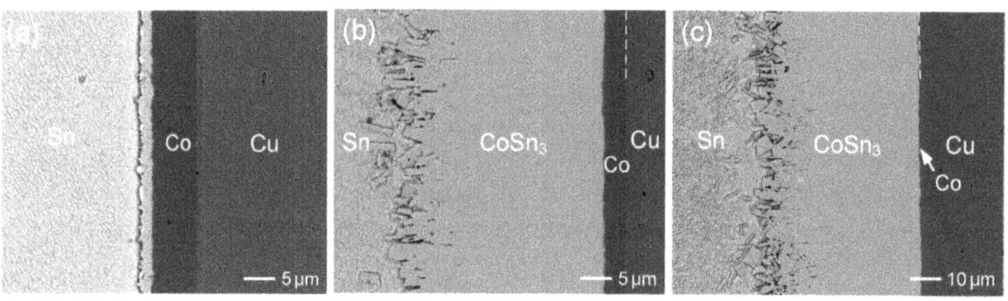

Figure 6. BEI micrographs of the Sn/Co reaction at 250 °C for (**a**) 10 s, (**b**) 1 h, and (**c**) 2 h.

For the Sn/Co-0.55Zn reaction, as the reaction time was extended to 1.5 h, the Co-0.55Zn layer was completely depleted, and the formed $CoSn_3$ layer detached from the Cu substrate, as shown in Figure 5e. EPMA analysis revealed that the bright phase of the detached layer had a composition of Sn-24.3 at.%Co-1.0 at.%Cu-0.1 at.%Zn, corresponding to the $CoSn_3$ phase. The adjacent dark phase exhibited a composition of Sn-10.7 at.%Co-42.7

at.%Cu-0.02 at.% Zn and was identified as the $(Cu,Co)_6Sn_5$ phase. On the Cu substrate side, the dark phase consisted of numerous Cu_6Sn_5 grains with a composition of Sn-2.1 at.%Co-46.7 at.%Cu-0.13 at.%Zn. Since the Co content in these grains was relatively low, they were designated as Cu_6Sn_5 to distinguish them from the dispersed $(Cu,Co)_6Sn_5$ phase with higher Co solubility. Figure 5f shows a magnified view of the $(Cu,Co)_6Sn_5$ phase and the porous Cu_6Sn_5 phase. The $(Cu,Co)_6Sn_5$ phase exhibited a striped morphology, with the $CoSn_3$ phase surrounded by $(Cu,Co)_6Sn_5$. This observation suggested that the formation of the $(Cu,Co)_6Sn_5$ phase likely occurred via a transformation of $CoSn_3$, induced by Cu dissolution from the substrate. Additionally, it is noteworthy that the porous Cu phase maintained a stable layered morphology. The grains remained intact and did not disintegrate or spall into the solder.

With an increase in Zn content to 2.5 wt.%, the interfacial microstructures were examined. As shown in Figure 7a, the initial interfacial structure resembled that of the Co-0.55Zn reaction. The Co-2.5Zn deposit was unevenly consumed due to dissolution, resulting in a rough and irregular interface. As the reaction progressed, the $CoSn_3$ layer thickness increased to approximately 13 μm after 10 min of aging and to around 32 μm after 30 min (Figure S1). The $CoSn_3$ phase contained a minor amount of 0.26 at.%Zn. As shown in Figure 7b, the thickness of the $CoSn_3$ layer after 1 h of aging was comparable to that observed in the Co-0.55Zn reaction.

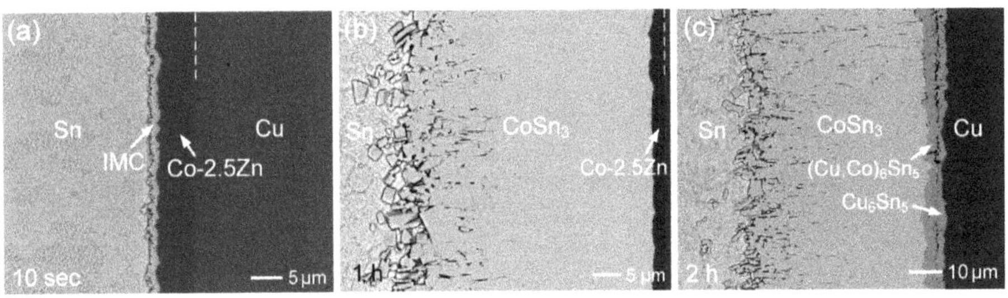

Figure 7. BEI micrographs of the Sn/Co-2.5Zn reaction at 250 °C for (**a**) 10 s, (**b**) 1 h, and (**c**) 2 h.

After reacting for 2 h (Figure 7c), the deposited layer was completely consumed, and the Cu substrate began to participate in the reaction. EPMA analysis, as shown in Table S1, indicated that the resulting reaction phases were $CoSn_3$, $(Cu,Co)_6Sn_5$, and Cu_6Sn_5. A crack was observed at the interface between the $(Cu,Co)_6Sn_5$ and Cu_6Sn_5 phases, leading to the detachment of the reaction layer from the substrate. Furthermore, as shown in Figures 8 and S2, the Co-4.8Zn deposit also exhibited similar interfacial behavior and a comparable trend in IMC growth rate to that observed in the Co-2.5Zn reaction. In all three Co-Zn coatings, Sn atoms from the solder diffused through the $CoSn_3$ layer and subsequently reacted with the Co-Zn substrate to form a new $CoSn_3$ phase.

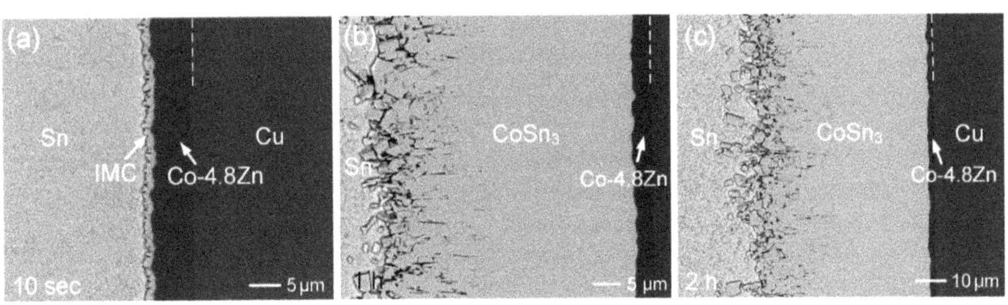

Figure 8. BEI micrographs of the Sn/Co-4.8Zn reaction at 250 °C for (**a**) 10 s, (**b**) 1 h, and (**c**) 2 h.

For the Zn content of 8.8 wt.%, the interfacial morphologies at various reaction durations are shown in Figure 9a–e. After aging for 10 min and 30 min, the IMC thickness increased to approximately 6 μm and 7.5 μm, respectively. These values were significantly lower compared to the IMC growth observed in the other three Co-Zn coatings. This suggests that the 8.8 wt.%Zn content was effective in suppressing IMC growth by ~50%, compared to the Sn/Co reaction. The prior study [27] reported that even a low Zn concentration of 0.1 wt.% in Sn solder could significantly inhibit $CoSn_3$ growth by 43%, and that increasing the Zn content to 0.5 wt.% further enhanced the suppression effect to 76%. The minor Zn addition in solder was suggested to inhibit the chemical reaction responsible for the nucleation and formation of $CoSn_3$. Consequently, the dissolution of the Co-8.8Zn deposit into the solder likely increased the Zn concentration, the dissolution of the Co-8.8Zn deposit into the solder likely increased the Zn concentration, leading to further suppression of IMC growth. Additionally, EPMA analysis in Figure 9b revealed that the $CoSn_3$ phase contained approximately 0.45 at.%Zn, which aligns with the findings from the prior study [27]. Although the Zn concentration in the $CoSn_3$ phase was slightly higher than that observed in the other three Co-Zn coatings, it remained at a very low level, suggesting that most of the Zn from the Co-Zn deposits dissolved into the solder.

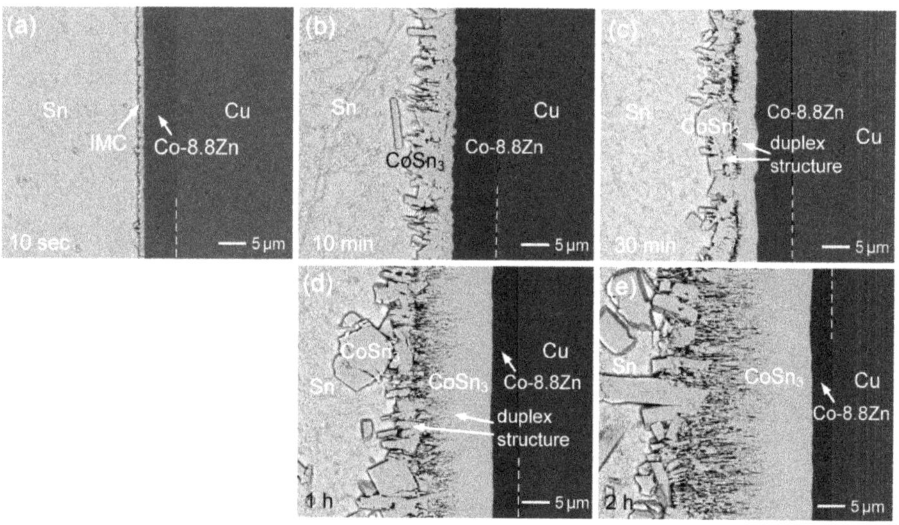

Figure 9. BEI micrographs of the Sn/Co-8.8Zn reaction at 250 °C for (**a**) 10 s, (**b**) 10 min, (**c**) 30 min, (**d**) 1 h, and (**e**) 2 h.

To better understand the influence of Zn dissolution from the Co-Zn coating into the solder, a simplified estimation was performed. Assuming that a 1 μm thick Co-8.8 wt.%Zn layer dissolves into a 100 μm thick interfacial region of molten Sn, the resulting average Zn concentration in the solder is calculated to be approximately 0.107 wt.%. Although this concentration is relatively low, it may still affect the nucleation and growth behavior of IMC at the interface. This estimation is consistent with previous findings [27], where even small additions of Zn (~0.1 wt.%) have been reported to suppress IMC growth. Furthermore, the suppression of excessive $CoSn_3$ growth in Co-Zn coatings can be primarily attributed to the influence of Zn on the nucleation behavior of $CoSn_3$. During the interfacial reaction, Sn acts as the dominant diffusing species, and the growth front of $CoSn_3$ is located at the interface between the $CoSn_3$ layer and the underlying Co-Zn coating. Zn atoms, either incorporated within the coating or segregated at the interface, can alter the interfacial energy and local chemical environment, increasing the nucleation energy barrier for $CoSn_3$ formation. As a result, the number of active nucleation sites is reduced and the initial formation rate of $CoSn_3$ is slowed, which further suppresses the overall IMC growth kinetics.

As presented in Figure 9b,c, the IMC phase displayed a distinct duplex morphology, with two distinct microstructures separated by a crack. This implies that different formation mechanism governed the development of the IMC phase. The inner $CoSn_3$ layer, formed adjacent to the Co-8.8 wt.%Zn deposit, was formed through the diffusion of Sn and its subsequent reaction with the Co-Zn coating at the interface. However, its growth was notably suppressed, especially during the initial reaction stage. On the other side, the IMC phase near the solder exhibited large grains, likely formed through the dissolution of the underlying $CoSn_3$ layer, followed by reprecipitation and grain growth. The crack separating the inner and outer $CoSn_3$ layers is likely induced by growth stress generated during the formation of the inner layer at the interface between the Co-Zn coating and the IMC. The volumetric expansion associated with $CoSn_3$ formation, combined with constrained growth conditions, can lead to the accumulation of internal stress. When this stress exceeds the local mechanical strength, it may result in interfacial cracking or delamination between the two IMC sublayers. After prolonged reaction times of 1 h and 2 h, as shown in Figure 9d,e, respectively, the microstructure remained unchanged. However, the $CoSn_3$ phase on the solder side exhibited significantly larger faceted grains, while the inner $CoSn_3$ layer, which displayed a striped grain structure, showed substantial growth. In the later stage of the reaction, the suppression effect significantly decreased, and the growth of the inner $CoSn_3$ phase became dominant.

The Co-Zn coatings demonstrated good adhesion to the Cu substrate across all Zn contents during the experimental process. Even after high-temperature interfacial reactions, quenching, and subsequent metallographic preparation, no obvious delamination was observed. Typically, higher doping levels (e.g., increased Zn content), elevated plating current densities, and thicker deposits can induce greater internal stress within the coating, which may compromise adhesion and lead to peeling. Therefore, special attention must be paid to controlling internal stress during the coating fabrication process. In Figure 9c, a seam resembling peeling was observed at the interface between the Co-8.8Zn deposit and the Cu substrate. However, this feature is considered a superficial surface artifact rather than a true interfacial separation. A more detailed discussion is provided in Figure S3 of the Supplementary File.

The thermal expansion mismatch was further analyzed. Table 3 summarizes the coefficients of thermal expansion (CTE) of the relevant materials. The mismatch between the Cu substrate and the Co-Zn coatings is relatively minor and unlikely to induce significant thermal stress. In contrast, a more substantial mismatch exists between the Sn solder and the Co-Zn/Cu substrate, which may generate considerable thermal stress during cooling from the soldering temperature. Notably, the CTE of Co-Zn alloys is lower than that of Sn and closer to that of Cu, potentially helping to mitigate thermal strain at the solder/coating interface. Despite this inherent thermal mismatch, no evident delamination was observed, indicating that the Co-Zn coatings maintained good mechanical integrity throughout the soldering process.

Table 3. Coefficients of thermal expansion (CTE) of relevant materials [35]. The CTE of Co-Zn alloys is not reported in literature. Based on the rule of mixtures and the known CTE values of pure Co and Zn, the CTE of Co-Zn alloys with Zn content below 10 wt.% is estimated.

Material	CTE ($\times 10^{-6}$ K^{-1})
Co	~13
Cu	~16.5
Sn (β-Sn)	~22
Zn	30.2
Co-Zn alloys (estimated)	~14–16 (depending on Zn content)

Figure 10 shows the IMC thickness as a function of reaction time for various Co-Zn deposits. For the Co-0.55Zn, Co-2.5Zn, and Co-4.8Zn samples, the IMC growth rate was slightly higher than that observed for pure Co. This unexpected trend could be attributed to the insufficient Zn content in these alloys, which was likely inadequate to form a continuous or effective barrier at the interface. At low Zn concentrations (e.g., 0.55–4.8 wt.%), Zn likely promotes the initial dissolution of the Co-Zn deposits into the molten Sn, thereby increasing the local availability of Co atoms and accelerating the formation of the $CoSn_3$ phase. In this regime, the Zn content is insufficient to significantly alter the interfacial chemistry or form a continuous Zn-rich barrier layer. As a result, the dominant growth mechanism remains similar to that of pure Co, and the slight enhancement in IMC thickness can be attributed to the more efficient supply of Co into the solder matrix. Conversely, the Co-8.8Zn alloy exhibited significantly suppressed IMC growth. This may result from the higher Zn content altering the interfacial chemistry, which hinders the nucleation of $CoSn_3$. Moreover, high Zn content may have promoted the dissolution and subsequent reprecipitation of $CoSn_3$ into coarser grains, leading to a slower net growth rate of the IMC layer. These results indicate that only when the Zn content exceeds a certain threshold (e.g., 8.8 wt.%) does Zn begin to exert a pronounced suppression effect.

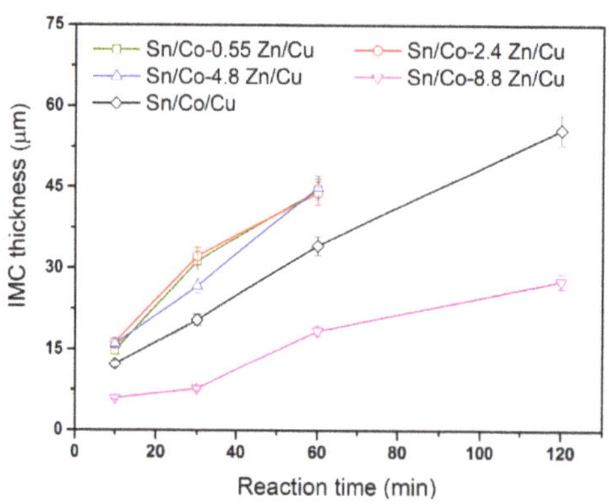

Figure 10. IMC thickness as a function of reaction time in the reactions between Sn and Co-Zn deposits.

In addition to Zn, other alloying strategies have also been investigated for suppressing interfacial IMC growth. Electroless Co-P coatings, typically deposited using sodium hypophosphite as a reducing agent, usually contain high phosphorus contents (~6 wt.%), which have been reported to accelerate interfacial reactions rather than suppress them. In contrast, electroplated Co-P coatings with a low phosphorus content (e.g., 0.7 wt.%) have been shown to effectively suppress IMC growth [18]. Notably, the effect of P doping exhibits an opposite trend compared to that of Zn in the Co-Zn system, where a higher Zn content leads to a stronger suppression effect.

3.3. Interfacial Reactions of Co-Zn Deposits with Sn at 160 °C

The solid-state interfacial reactions between Sn and various Co-Zn deposits were further investigated at 160 °C. Figure 11a–e present the interfacial microstructures after aging for 120 h. Both the Co-0.55Zn and Co-2.5Zn samples exhibited similar $CoSn_3$ layer thicknesses of approximately 11 μm, which were significantly thinner than that of the pure Co sample (~20 μm). This indicates that, in the solid-state reaction, even small amounts of Zn in the Co deposit were effective in hindering the formation of IMCs. Moreover, the suppression effect became more pronounced with increasing Zn content. The Co-4.8Zn sample exhibited an even thinner IMC layer, measuring only ~5 μm, representing a reduction of approximately 75% compared to pure Co. This suggests that Zn likely affects the nucleation or growth rate of $CoSn_3$ and possibly alters the diffusion kinetics or the interaction of Sn with the Co-Zn deposits.

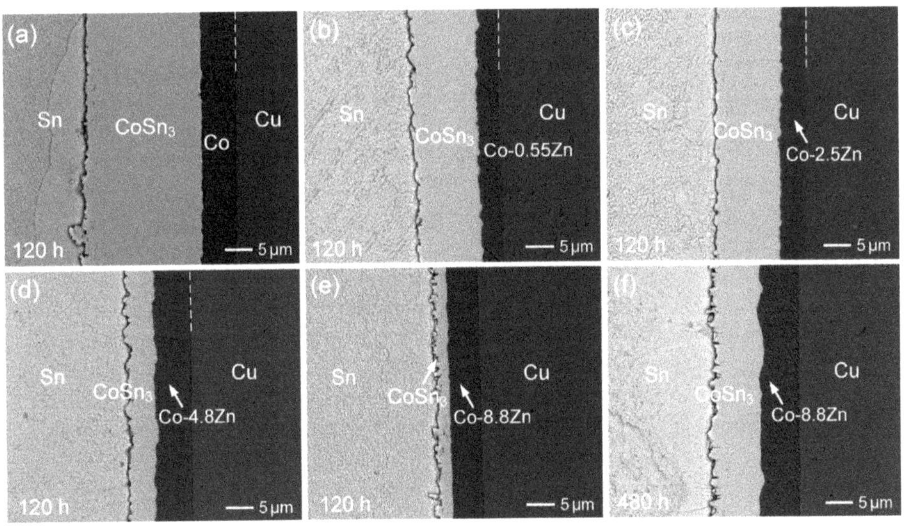

Figure 11. BEI micrographs of the interfacial reactions between Sn and various Co-Zn deposits after aging at 160 °C for 120 h: (**a**) Co, (**b**) Co-0.55Zn, (**c**) Co-2.5Zn, (**d**) Co-4.8Zn, and (**e**) Co-8.8Zn. (**f**) Co-8.8Zn after 480 h.

Figure 11e further confirms this trend, showing that the Co-8.8Zn sample exhibited a significantly reduced IMC thickness of approximately 2.5 μm after aging for 120 h. Even after extended aging for 480 h, the IMC layer grew to only around 9 μm, indicating that the suppression remained effective. These results clearly demonstrate that the inhibition of IMC growth in the solid-state reaction is considerably more pronounced than in the liquid-state reaction. This may be attributed to the slower diffusion rates at lower temperatures, which provide more time for Zn to affect the interfacial reaction. Additionally, these findings suggest that optimizing the Zn content in Co-Zn alloys could be a promising strategy for controlling IMC formation in soldering processes, especially for applications that require long-term reliability.

This study primarily focuses on the fabrication of Co-Zn deposits and the interfacial reaction behavior of Co-Zn coatings during the liquid-state reflow process. Compared to conventional Ni, Co, and Co-P systems, the Co-Zn alloy exhibits unique alloying effects and enhanced interfacial stability. The Co-Zn system not only maintains excellent barrier properties and thermal stability but also offers the potential to modulate interfacial reaction kinetics through Zn addition, effectively suppressing the formation of detrimental phases. Furthermore, Co-Zn alloys demonstrate promising diffusion barrier performance and solder compatibility, which can contribute to improved reliability in electronic packaging interconnects.

To fully evaluate the practical applicability of these coatings, future work will include long-term aging experiments and mechanical reliability assessments of solder joints with varying Zn contents. Such investigations are essential for verifying the durability and performance of Co-Zn coatings under practical service conditions, thereby further supporting their potential for industrial applications in microelectronic packaging. Moreover, further studies are necessary to elucidate the fundamental mechanisms by which Zn influences interfacial reactions, particularly its effects on diffusion behavior and phase formation.

4. Conclusions

This study successfully fabricated Co-Zn alloy coatings with varying Zn contents via the electroplating method. The compositional trends of the coatings exhibited an anomalous co-deposition behavior, where the less noble Zn had a greater tendency to deposit than Co. Increasing Zn content significantly altered the surface morphologies and cross-sectional microstructures, resulting in finer grains and a transition from columnar to dendritic structures. XRD analysis revealed that Zn incorporation reduced the crystallinity of the Co phase and induced partial amorphization. Interfacial reaction studies with Sn solder at 250 °C and 160 °C demonstrated the critical role of Zn in IMC formation. At 250 °C, lower Zn contents (0.55–4.8 wt.%) slightly accelerated $CoSn_3$ growth due to enhanced dissolution of the Co-Zn layer. However, higher Zn content (8.8 wt.%) significantly suppressed IMC growth by approximately 50%. Additionally, the $CoSn_3$ phase exhibited a dense, layered structure for Zn contents below 4.8 wt.%. In contrast, the $CoSn_3$ in the Co-8.8Zn reaction displayed a duplex structure with two distinct microstructures. In the solid-state reactions at 160 °C, the suppression effect of Zn was more pronounced. The addition of 0.55 wt.% Zn significantly reduced the IMC growth. As the Zn content increased, the suppression effect became more evident. The Co-8.8Zn sample maintained a thin IMC layer, showing over 85% suppression, even after long-term aging. These findings demonstrate that Zn plays a key role in modifying interfacial reactivity and microstructure evolution. Co-Zn coatings with higher Zn content effectively suppress excessive IMC growth, which is crucial for enhancing the thermal and mechanical reliability of solder joints. This makes them promising candidates for surface finishes in microelectronic packaging and advanced lead-free soldering applications.

Supplementary Materials: The following supporting information can be downloaded at: https://www.mdpi.com/article/10.3390/ma18122680/s1, Figure S1: BEI micrographs of the Sn/Co-2.5Zn reaction at 250 °C for (a) 10 min and (b) 30 min; Figure S2: BEI micrographs of the Sn/Co-4.8Zn reaction at 250 °C for (a) 10 min and (b) 30 min; Figure S3: BEI micrographs of the Sn/Co-8.8Zn reaction at 250 °C for (a) 10 sec, (b) 10 min, (c) 30 min, and (b) 30 min; Figure S4: Cross-sectional micrograph of the Sn/Co–2.5Zn/Cu sample reacted at 250 °C for 2 h (identical to Figure 7c in the main text), provided here for reference and further analysis; Table S1: EPMA analysis in Sn/Co-2.5Zn/Cu reaction at 250 °C for 2 h. The analyzed positions were shown in the figure below.

Author Contributions: Conceptualization, C.-H.W.; data curation, C.-H.W. and C.-Y.L.; funding acquisition, C.-H.W.; investigation, C.-Y.L.; methodology, C.-H.W. and C.-Y.L.; project administration, C.-H.W.; supervision, C.-H.W.; writing—original draft, C.-H.W.; writing—review and editing, C.-H.W. All authors have read and agreed to the published version of the manuscript.

Funding: This research was funded by National Science and Technology Council of Taiwan, R.O.C. (grant number NSTC 113-2221-E-194-003).

Institutional Review Board Statement: Not applicable.

Informed Consent Statement: Not applicable.

Data Availability Statement: The original contributions presented in this study are included in the article/Supplementary Material. Further inquiries can be directed to the corresponding author.

Conflicts of Interest: The authors declare no conflict of interest.

References

1. Dele-Afolabi, T.T.; Ansari, M.N.M.; Azmah Hanim, M.A.; Oyekanmi, A.A.; Ojo-Kupoluyi, O.J.; Atiqah, A. Recent advances in Sn-based lead-free solder interconnects for microelectronics packaging: Materials and technologies. *J. Mater. Res. Technol.* **2023**, *25*, 4231–4263. [CrossRef]
2. Cheng, S.; Huang, C.M.; Pecht, M. A review of lead-free solders for electronics applications. *Microelectron. Reliab.* **2017**, *75*, 77–95. [CrossRef]
3. Ma, H.; Suhling, J.C. A review of mechanical properties of lead-free solders for electronic packaging. *J. Mater. Sci.* **2009**, *44*, 1141–1158. [CrossRef]
4. Sun, Q.; Wang, J.; Wang, X.-N.; Zhou, X.-W.; Tang, X.-X.; Akira, K. Cu aggregation behavior on interfacial reaction of Sn-3.0Ag-0.5Cu/ENIG solder joints. *Mater. Lett.* **2023**, *348*, 134659. [CrossRef]
5. Kim, J.; Jung, S.-B.; Yoon, J.-W. Effect of Ni(P) thickness in Au/Pd/Ni(P) surface finish on the electrical reliability of Sn–3.0Ag–0.5Cu solder joints during current-stressing. *J. Alloys Compd.* **2015**, *622*, 529–534. [CrossRef]
6. Choi, H.; Kim, C.-L.; Sohn, Y. Diffusion barrier properties of the intermetallic compound layers formed in the Pt nanoparticles alloyed Sn-58Bi solder joints reacted with ENIG and ENEPIG surface finishes. *Materials* **2022**, *15*, 8419. [CrossRef]
7. Wang, J.Y.; Tang, Y.K.; Yeh, C.Y.; Chang, P.J.; Lin, Y.X.; Lin, E.J.; Wu, C.Y.; Zhuang, W.X.; Liu, C.Y. Kinetics of Ni solid-state dissolution in Sn and Sn3.5Ag alloys. *J. Alloys Compd.* **2019**, *797*, 684–691. [CrossRef]
8. Huang, L.C.; Zhang, Y.P.; Chen, C.M.; Hung, L.Y.; Wang, Y.P. Interfacial reactions between pure indium solder and Au/Ni metallization. *J. Mater. Sci. Mater. Electron.* **2022**, *33*, 13143–13151. [CrossRef]
9. Wang, C.-H.; Chen, H.-H.; Li, P.-Y.; Chu, P.-Y. Kinetic analysis of Ni_5Zn_{21} growth at the interface between Sn-Zn solders and Ni. *Intermetallics* **2012**, *22*, 166–175. [CrossRef]
10. Yao, J.; Li, C.; Shang, M.; Chen, X.; Wang, Y.; Ma, H.; Ma, H.; Liu, X. Diffusion barrier performance of Ni-W layer at Sn/Cu interfacial reaction. *Materials* **2024**, *17*, 3682. [CrossRef]
11. Liu, L.; Shi, L.; Peng, J.; Jiang, B.; Liu, S.; Liu, C.; Chen, Z. Interfacial reaction between Sn–Ag solder and electroless Ni–Fe–P diffusion barriers with different internal microstructure. *Mater. Res. Bull.* **2022**, *152*, 111854. [CrossRef]
12. Xiao, J.; Wang, F.; Li, J.; Chen, Z. Comparison of interfacial reactions and isothermal aging of cone Ni–P and flat Ni–P with Sn3.5Ag solders. *Appl. Surf. Sci.* **2023**, *625*, 157219. [CrossRef]
13. Humpston, G. Cobalt: A universal barrier metal for solderable under bump metallisations. *J. Mater. Sci. Mater. Electron.* **2010**, *21*, 584–588. [CrossRef]
14. Bi, X.; Hu, X.; Jiang, X.; Li, Y. Mechanical properties of $CoSn_2$ and α-$CoSn_3$ intermetallic compounds: First-principles calculations and nano-indentation measurements. *Appl. Phys. A* **2019**, *125*, 217. [CrossRef]
15. Liu, Y.; Li, C.; Chen, P.; Liu, J.; Hu, A.; Li, M. Effect of Co-W and Co-Fe-W diffusion barriers on the reliability of the solder/Cu interface during reflow conditions. *Electron. Mater. Lett.* **2024**, *20*, 517–524. [CrossRef]
16. Wang, C.-H.; Kuo, C.Y. Growth Kinetics of the solid-state interfacial reactions in the Sn–Cu/Co and Sn/Co–Cu couples. *Mater. Chem. Phys.* **2011**, *130*, 651–656. [CrossRef]
17. Pan, H.-C.; Hsieh, T.-E. Diffusion barrier characteristics of electroless Co(W,P) thin film to lead-free Sn-Ag-Cu solder. *J. Electrochem. Soc.* **2011**, *158*, 123–129. [CrossRef]
18. Wang, C.-H.; Huang, S.-E.; Chiu, C.-W. Influence of the P content on phase formation in the interfacial reactions between Sn and electroless Co(P) metallization on Cu substrate. *J. Alloys Compd.* **2015**, *619*, 474–480. [CrossRef]
19. Lu, N.; Yang, D.; Li, L. Interfacial reaction between Sn-Ag-Cu solder and Co-P films with various microstructures. *Acta Mater.* **2013**, *61*, 4581–4590. [CrossRef]
20. Nishizawa, K.; Matsumoto, A.; Nakagawa, Y.; Sakuma, H.; Goto, S.; Fukumuro, N.; Yae, S. Comparison of electroless Ni-P and Co-W-P diffusion properties against GaAs substrate. *J. Electron. Mater.* **2023**, *52*, 4080–4090. [CrossRef]
21. Tian, S.; Zhou, J.; Xue, F.; Cao, R.; Wang, F. Microstructure, interfacial reactions and mechanical properties of Co/Sn/Co and Cu/Sn/Cu joints produced by transient liquid phase bonding. *J. Mater. Sci. Mater. Electron.* **2018**, *29*, 16388–16400. [CrossRef]
22. Wang, C.-H.; Hsieh, H.-C.; Lee, H.-Y.; Wu, A.-T. Co-P diffusion barrier for p-Bi_2Te_3 thermoelectric material. *J. Electron. Mater.* **2019**, *48*, 53–57. [CrossRef]

23. Zhen, C.; Ma, L.; Liu, S.; Wang, Y.; Li, D.; Guo, F. Effect of P content on diffusion resistance and interfacial mechanical properties of crystalline Co–P coatings in solder joints. *J. Mater. Sci. Mater. Electron.* **2023**, *34*, 333. [CrossRef]
24. Wang, C.-H.; Kuo, C.-Y.; Huang, S.-E.; Li, P.-Y. Temperature effects on liquid-state Sn/Co interfacial reactions. *Intermetallics* **2013**, *32*, 57–63. [CrossRef]
25. Wang, C.-H.; Chen, S.-W. Sn/Co solid/solid interfacial reactions. *Intermetallics* **2008**, *16*, 524–530. [CrossRef]
26. Hong, J.H.; Lee, H.Y.; Wu, A.T. Massive spalling and morphological change of intermetallic compound affected by adding Pd in Co-based surface finishes. *J. Alloys Compd.* **2013**, *580*, 195–200. [CrossRef]
27. Wang, C.-H.; Huang, S.-E.; Li, K.-T. Inhibiting $CoSn_3$ growth at the Sn/Co system by minor Zn addition. *Intermetallics* **2015**, *56*, 68–74. [CrossRef]
28. Wang, C.-H.; Kuo, C.-Y.; Guo, Y.-B. Effects of minor Cu, Ni and Ag additions on the reactions between Sn-based solders and Co substrate. *JOM* **2019**, *71*, 3023–3030. [CrossRef]
29. Lodhi, Z.F.; Mol, J.M.C.; Hamer, W.J.; Terryn, H.A.; De Wit, J.H.W. Cathodic inhibition and anomalous electrodeposition of Zn–Co alloys. *Electrochim. Acta* **2007**, *52*, 5444–5452. [CrossRef]
30. Zhang, S.; Yu, J.; Liu, Z.; Yin, Y.; Qiao, C. Numerical and experimental investigation of the effect of current density on the anomalous codeposition of ternary Fe-Co-Ni alloy coatings. *Materials* **2022**, *15*, 6141. [CrossRef]
31. Pandiyarajan, S.; Ganesan, M.; Liao, A.-H.; Manickaraj, S.S.M.; Huang, S.T.; Chuang, H.-C. Ultrasonic-assisted supercritical-CO_2 electrodeposition of Zn-Co film for high-performance corrosion inhibition: A greener approach. *Ultrason. Sonochem.* **2021**, *72*, 105463. [CrossRef] [PubMed]
32. Yunus, M.; Capel-Boute, C.; Decroly, C. Inhibition effect of zinc on the cathodic deposition of cobalt—I. Electrochemical and structural observations in sulphate solutions. *Electrochim. Acta* **1965**, *10*, 885–894. [CrossRef]
33. Higashi, K.; Fukushima, H.; Urakawa, T.; Adaniya, T.; Matsudo, K. Mechanism of the electrodeposition of zinc alloys containing a small amount of cobalt. *J. Electrochem. Soc.* **1981**, *128*, 2081–2085. [CrossRef]
34. Abou-Krisha, M.M. Electrochemical studies of zinc–nickel codeposition in sulphate bath. *Appl. Surf. Sci.* **2005**, *252*, 1035–1048. [CrossRef]
35. WebElements. Available online: https://www.webelements.com (accessed on 28 May 2025).

materials

MDPI

Article

Determination of Strength Parameters of Composite Reinforcement Consisting of Steel Member, Adhesive, and Carbon Fiber Textile

Maciej Adam Dybizbański [1,*], Katarzyna Rzeszut [1], Saydiolimkhon Abdusattarkhuja [1] and Zheng Li [2]

1 Faculty of Civil and Transport Engineering, Poznań University of Technology, 5 Marii Skłodowskiej-Curie Str., 60-965 Poznań, Poland; katarzyna.rzeszut@put.poznan.pl (K.R.); saydiolimkhon.abdusattarkhuja@put.poznan.pl (S.A.)

2 Institute of Civil Engineering, Technische Universität Berlin, G.-Meyer-Alle 25, 13355 Berlin, Germany; dr.zheng.li@web.de

* Correspondence: maciej.dybizbanski@doctorate.put.poznan.pl

Abstract: The main aim of the study was the determination of the strength parameters of composite bonded joints consisting of galvanised steel elements, an adhesive layer, and Carbon-Fiber-Reinforced Plastic (CFRP) fabric. For this purpose, shear laboratory tests were carried out on 60 lapped specimens composed of 2 mm thick hot-dip galvanised steel plates of S350 GD. The specimens were overlapped on one side with SikaWrap 230 C carbon fibre textile (CFT) using SikaDur 330 adhesive. The tests were carried out in three series that differed in overlap length (15 mm, 25 mm, and 35 mm). A discussion on the failure mechanism in the context of the bonding capacity of the composite joint was carried out. We observed three forms of joint damage, namely, at the steel-adhesive interface, fibre rupture, and mixed damage behaviour. Moreover, an advanced numerical model using the commercial finite element (FE) program ABAQUS/Standard and the coupled cohesive zone model was developed. The material behaviour of the textile was defined as elastic-lamina and the mixed-mode Hashin damage model was implemented with bi-linear behaviour. Special attention was focused on the formulation of reliable methodologies to determine the load-bearing capacity, failure mechanisms, stress distribution, and the strength characteristics of a composite adhesive joint. In order to develop a reliable model, validation and verification were carried out and self-correlation parameters, which brought the model closer to the laboratory test, were proposed by the authors. Based on the conducted analysis, the strength characteristics including the load-bearing capacity, failure mechanisms, and stress distribution were established. The three forms of joint damage were observed as steel-adhesive interface failure, fibre rupture, and mixed-damage behaviour. Complex interactions between the materials were observed. The most dangerous adhesive failure was detected at the steel and adhesive interface. It was also found that an increase in adhesive thickness caused a decrease in joint strength. In the numerical analysis, two mechanical models were employed, namely, a sophisticated model of adhesive and fabric components. It was found that the fabric model was very sensitive to the density of the finite element mesh. It was also noticed that the numerical model referring to the adhesive layer was nonsensitive to the mesh size; thus, it was regarded as appropriate. Nevertheless, in order to increase the reliability of the numerical model, the authors proposed their own correlation coefficients α and β, which allowed for the correct mapping of adhesive damage.

Keywords: adhesive joints; composite fabric; cold-formed steel elements; numerical simulation; cohesive zone model; laboratory tests

1. Introduction

The phenomenon of adhesion is fundamental to many applications across scientific disciplines, from the development of advanced adhesives in engineering to the evolutionary adaptations seen in nature. Understanding the intricate interplay in adhesion forces is crucial for optimising these applications and enhancing material performance [1]. The complexity in adhesion mechanisms not only highlights the importance of physicochemical interactions but also underscores the diversity that strategies organisms employ to navigate their environments [2].

A profound comprehension of adhesion forces and their underlying physical properties might provide a suitable framework for the design of bonded joints [3]. Figure 1 represents the most notable and common physicochemical forces found in nature and illustrates their interactions between two surfaces, which used to be called adhesion. The interpretation of the term "adhesion" differs across various domains of research, including chemistry, biology, physics, and engineering. Surface forces are the fundamental interactions that facilitate the convergence of two bodies, while adhesion refers to the energy expended to maintain their connection.

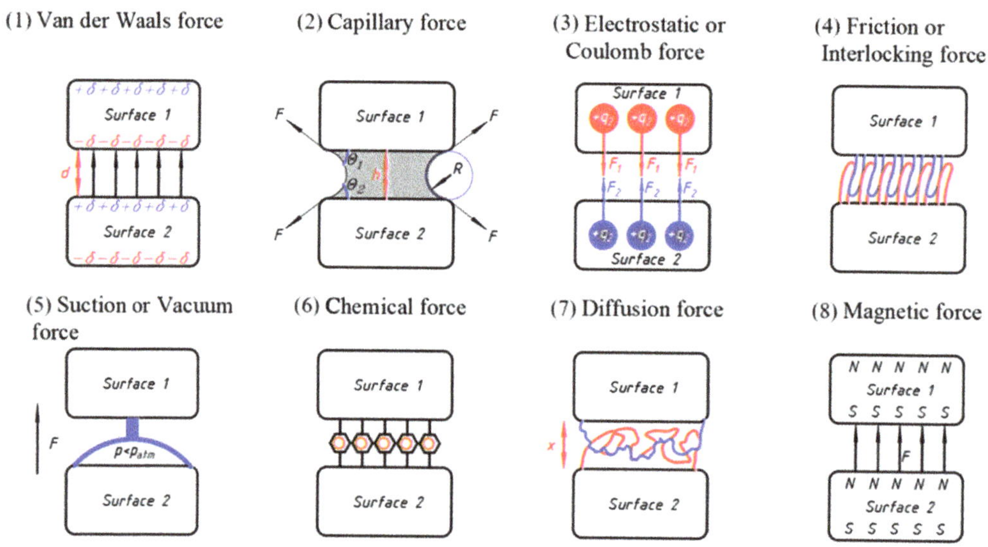

Figure 1. Schematic illustration of the mechanisms of adhesion and the various forces that exist between two surfaces in nature. (1) Van der Waals force; (2) capillary force; (3) electrostatic or coulomb force; (4) friction or interlocking force; (5) suction or vacuum force; (6) chemical force; (7) diffusion force; (8) magnetic force [3].

Figure 1 delivers a thorough and intricately detailed picture that captures a variety of adhesion mechanisms and forces, all of which are notably derived from the complexities of the natural ecosystem, with particular attention to the surprising adaptations and capabilities exhibited by different materials on slippery surfaces. Herein lie the crucial and salient details concerning the diverse adhesion mechanisms and the corresponding forces that are vividly depicted in the aforementioned figure:

Van der Waals forces—These are weak attractive forces that occur between molecules. In the context of adhesion, van der Waals forces play a significant role in the adhesion of geckos [4] and other creatures to surfaces. Figure 1 illustrates how these forces contribute to the overall adhesion when the animal's foot comes into contact with a surface [5].

In rigid materials, van der Waals forces contribute to adhesion by creating attractive interactions between surfaces, as demonstrated in the analysis of a rigid sphere and a half-elastic region [6]. In biological systems, van der Waals forces enhance cell adhesion, particularly in stem cells exposed to cold atmospheric plasma, which modifies the forces at play between cells and their extracellular matrix [7].

Capillary action, which can be defined as the ability of a liquid to flow in narrow spaces without the assistance of external forces, stands as a pivotal mechanism that significantly contributes to the phenomenon of adhesion, particularly under conditions characterised by the presence of moisture. The illustration provided serves to elucidate the manner in which the moisture present within the microstructures found in the feet of various animals, such as those exhibited by frogs, effectively enhances the process of adhesion through the engagement of capillary forces that operate at a microscopic level. This particular mechanism facilitates an improved ability to maintain grip on surfaces that are inherently slippery by drawing water into minuscule interstices, thereby substantially increasing the overall area of contact between the adhering surfaces [8].

Electrostatic forces also play an important role in adhesion. Figure 1 shows how positively charged amino acids in bristle surfaces promote adhesion through electrostatic interactions, enhancing their ability to adhere to different surfaces.

Friction or interlocking force refers to a type of adhesion where one surface adheres to another surface mechanically, increasing adhesion due to its physical structure. This structural adaptability allows animals living in appropriate natural conditions to move on slippery surfaces [9].

The suction force is generated when there is a pressure difference between the inside of a cavity (like a suction cup) and the external environment. This difference creates a vacuum effect that allows the suction cup to adhere to surfaces. The force can be mathematically expressed as follows:

$$F = p \cdot A, \tag{1}$$

where p is the pressure outside the suction cup (typically atmospheric pressure) and A is the area of the surface covered by the cup. This equation indicates that the suction force increases with the area of contact and the pressure differential.

The suction force works on the principle of creating a low-pressure zone inside a cavity. For instance, when an octopus uses its suckers, it contracts its muscles to expel liquid from the cavity, reducing the internal pressure and creating suction. This process allows the octopus to adhere to surfaces effectively [10].

Chemical forces arise from the interactions between molecules, primarily through bonds such as hydrogen bonds, covalent bonds, and ionic bonds. These forces are crucial for the adhesion between surfaces, influencing how materials stick together or resist separation.

Diffusion force refers to the movement of particles from an area of higher concentration to an area of lower concentration. This process is driven by the concentration gradient and is essential in various biological and physical systems, including adhesion mechanisms on slippery surfaces.

Magnetic force arises from the interaction between magnetic fields and charged particles. It can either attract or repel depending on the polarity of the magnetic charges involved. This force is crucial in various applications, including the design of materials that can enhance grip on slippery surfaces [11].

Speaking about the methods of investigating van der Waals forces, which are described above, we should note the measurement method shown in the article [12]. Advanced techniques such as interferometric force spectroscopy integrated with scanning electron microscopy allow for the measurement of van der Waals forces at sub-nanonewton resolutions, isolating these forces from other interactions [13].

In terms of the stress state of a steel plate, [14] outlines a significant advancement in assessing the mechanical stress state of steel plates using eddy currents and artificial intelligence (AI) techniques, aiming for high accuracy in stress mapping and evaluation.

It should be emphasised that the phenomenon of adhesion represents an exceedingly intricate, difficult, and complex problem. In engineering practices, it is postulated that the failure of a joint should not be of an adhesive nature, and in order to achieve that goal, it is needed to perform the appropriate preparation of the surfaces of the components. Reference [15] covers the developments in failure mechanics and the extended finite element method within the finite element system, supported by application examples. In addition, [15] presents classical methodologies of continuum mechanics and fracture mechanics and discusses boundary element and finite difference methods with an indication of their optimal applications. Also, [16] discusses the development of a stochastic model to analyse the behaviour of adhesive joints. This method aims to address the uncertainties in material properties and loading conditions that can affect the performance of these joints. By incorporating stochastic methods, the model can predict variability in the joint's response, providing a more comprehensive understanding of its behaviour under different conditions. This approach helps in designing more reliable and efficient adhesive joints for various engineering applications. Unfortunately, currently, there is no numerical method that universally solves all questions. However, the cohesive zone model (CZM) is useful for analysing adhesive joints. The widely recognised cohesive zone model (CZM) is predicated on the simulation of failure characterised by cohesion, specifically in scenarios where a fracture propagates within the adhesive layer. CZM has become the predominant methodology for investigating the dynamic behaviour of adhesive joints such as fatigue, variable strain rate, and impact. Although, in general, the dynamic behaviour of adhesive joints is divided into three areas: fatigue, variable strain rate and impact, and modal analysis [17]. Therefore, this model is very useful in the case of engineering analysis. Grounded in the principles of the CZM, the authors attempted to establish a model that accurately reflects adhesive failure at the adhesive–steel interface, which was also empirically observed in laboratory conditions.

The research presented in this paper concerns the formulation of reliable methodologies to determine the strength characteristics of a composite reinforcement consisting of the steel element, adhesive, and CFRP fabric. This issue of the investigation covers the characterisation and evaluation of an adhesive interface between galvanised steel and CFRP fabric with the aim of assessing its mechanical attributes, which include the load-bearing capacity of the bond, failure mechanisms, and stress distribution. Strength characteristics will be established through a synthesis of existing literature, numerical modelling, and empirical laboratory experimentation.

2. Experimental Tests

2.1. Problem Formulation

In the first part of the research, a lapped shear laboratory test was performed. The specimens were made from steel plates (2 mm thick) overlapped on one side with SikaWrap 230 C [18] carbon fibre textile (CFT) using SikaDur 330 [19] adhesive. Both SikaWrap 230 C and SikaDur 330 were provided by Sika Poland sp. z o.o. (Warsaw, Poland). The tests were carried out on 60 specimens, 3 series (A, B, and C) with 20 specimens each. For each series, a different overlap length was considered, namely 15 mm, 25 mm, and 35 mm (Table 1). The samples were produced as shown in Figure 2.

Table 1. Tested series.

No.	Series Label	Overlap Length (mm)
1	A	15
2	B	25
3	C	35

Before creating a bonding connection, the steel plates' thicknesses were measured. The steel plate surface was cleaned with sandpaper and degreased with acetone. After surface preparation, the adhesive layer was applied uniformly on the steel plate surface, followed by one layup of CFT. The CFT was pressed to the steel surface, creating a bonded connection and embedding the carbon fibres in the epoxy matrix. Moreover, the excess of adhesive was removed from the surface of the textile. After the hardening process of the adhesive was finished, which took 7 days, the measurement of each single lap joint

(SLJ) geometry was performed. The difference between the second and first measurements, reduced to a CFT thickness of 0.129 mm, was the adhesive thickness.

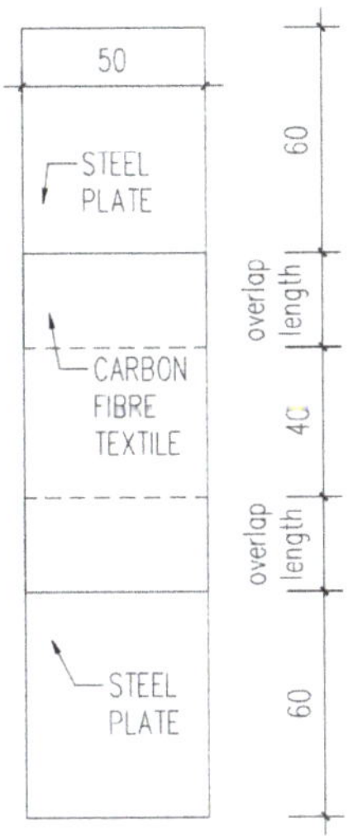

Figure 2. Schematic of the sample.

For experimental testing, an INSTRON SATEC 300 DX (Instron Polska, Opole, Poland) testing machine was used. The load accuracy measurement was 0.6 kN and the displacement accuracy was 0.13 mm. The testing machine was set to zero after the test loading system was assembled, but before the specimen was actually gripped at both ends. Once the force zero point was set, the force measurement system was not altered in any way during the test. The tensile tests were carried out under displacement control conditions. The samples were stretched at a speed of 0.01 mm/s until the specimens failed. An example of the prepared specimen in the grips of the testing machine is shown in Figure 3. After the completion of the laboratory tests, an observation of the nature of the fracture for all adhesive SLJs was conducted.

Figure 3. Specimen in the grips of a testing machine.

2.2. Experimental Test Results

The test results for series A, B and C are summarised in Tables 2–4. The mean value of the maximum load and the corresponding imposed displacement were reported for the respective series of the specimens. The standard deviation was determined for each parameter.

Table 2. Adhesive shear strength test results for the A series.

No.	Sample Label	Maximum Load (kN)	Displacement at Maximum Load (mm)
1	A1	5.60	2.25
2	A2	7.56	2.47
3	A3	4.94	2.04
4	A4	5.88	2.18
5	A5	7.84	2.49
6	A6	6.32	2.56
7	A7	7.74	2.43
8	A8	5.99	2.49
9	A9	8.87	2.75
10	A10	5.17	2.21
11	A11	7.42	2.58
12	A12	5.46	2.32
13	A13	4.91	2.18
14	A14	5.52	2.33
15	A15	5.15	2.26
16	A16	6.03	2.40
17	A17	3.48	2.11
18	A18	4.27	2.12
19	A19	7.91	2.87
20	A20	6.42	2.45
Average		6.12	2.37
Standard deviation		1.386	0.216

Table 3. Adhesive shear strength test results for the B series.

No.	Sample Label	Maximum Load (kN)	Displacement at Maximum Load (mm)
1	B1	5.46	2.16
2	B2	5.45	2.12
3	B3	3.84	1.82
4	B4	4.69	2.03
5	B5	6.14	2.30
6	B6	6.97	2.34
7	B7	6.94	2.28
8	B8	5.01	2.03
9	B9	4.75	1.85
10	B10	5.85	2.15
11	B11	4.50	1.98
12	B12	7.45	2.36
13	B13	7.47	2.33
14	B14	2.54	1.63
15	B15	5.70	2.05
16	B16	5.84	2.13
17	B17	4.84	2.04
18	B18	2.40	1.57
19	B19	3.95	1.86
20	B20	4.59	1.97
Average		5.22	2.05
Standard deviation		1.419	0.225

Diagrams of the force–displacement relationships for each series, A, B, and C, are shown in Figures 4–6, respectively.

Table 4. Adhesive shear strength test results for the C series.

No.	Sample Label	Maximum Load (kN)	Displacement at Maximum Load (mm)
1	C1	8.43	2.27
2	C2	9.98	2.44
3	C3	6.99	2.19
4	C4	7.93	2.29
5	C5	3.07	1.66
6	C6	6.41	2.17
7	C7	7.18	2.25
8	C8	7.10	2.17
9	C9	7.73	2.24
10	C10	5.07	1.90
11	C11	7.62	1.92
12	C12	7.84	2.34
13	C13	8.47	2.16
14	C14	9.55	2.48
15	C15	3.40	1.65
16	C16	9.06	2.59
17	C17	3.04	1.58
18	C18	6.35	2.13
19	C19	6.63	2.07
20	C20	6.30	2.03
Average		6.91	2.13
Standard deviation		1.992	0.273

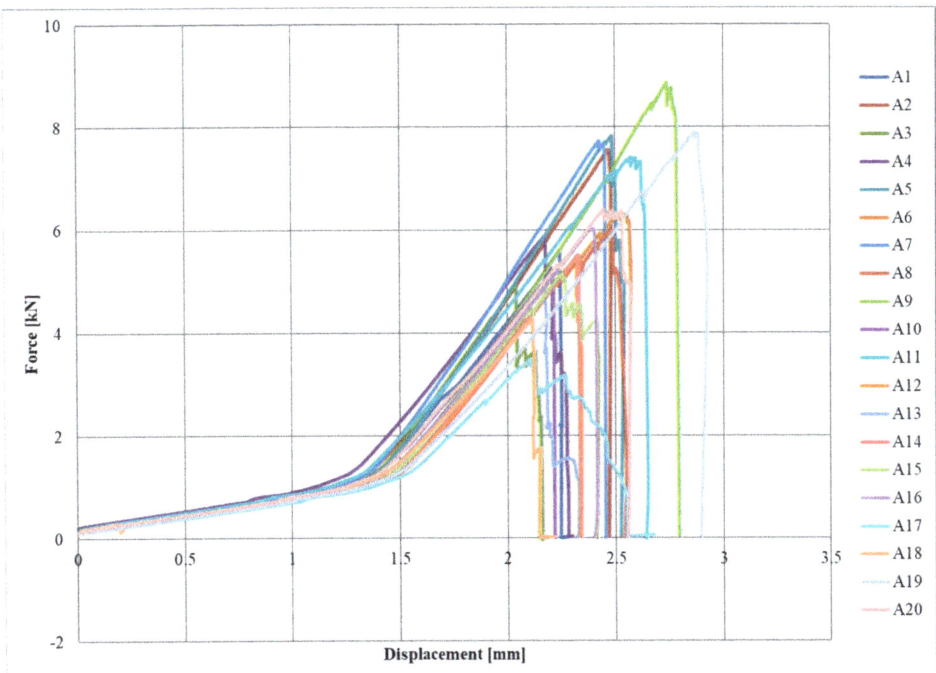

Figure 4. Force-displacement diagram for A series.

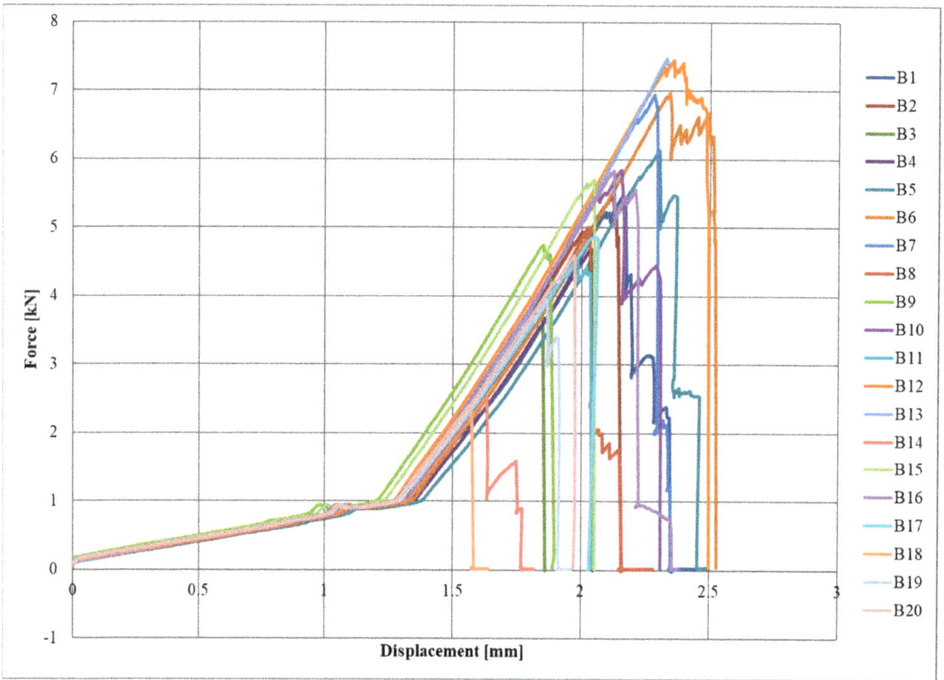

Figure 5. Force-displacement diagram for B series.

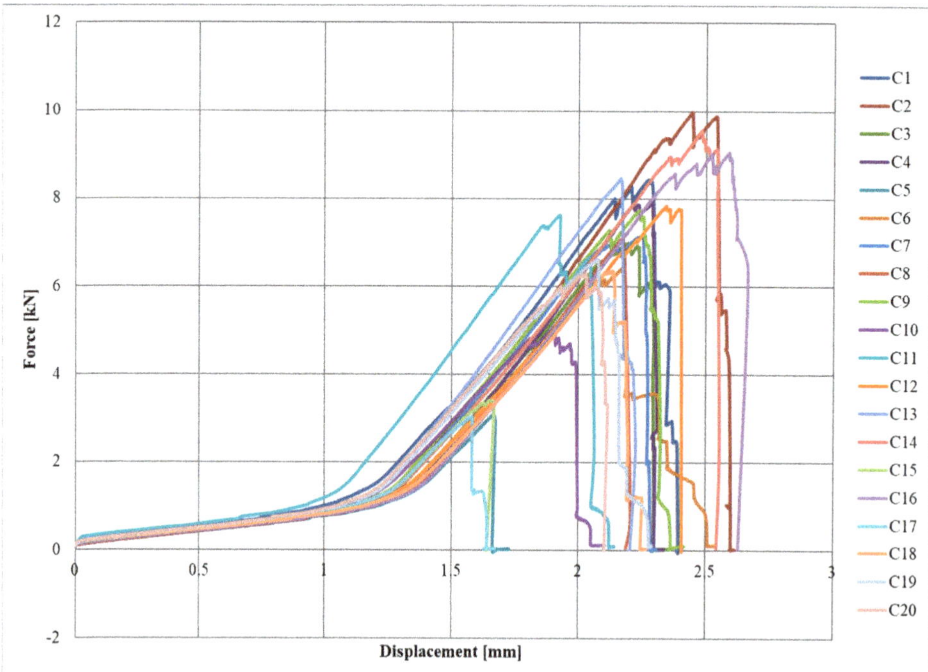

Figure 6. Force-displacement diagram for C series.

The initial slow increase in force due to the lower initial elastic modulus of the textile can be observed in the graphs. Once a displacement of approximately 1.1–1.5 mm was reached, there was a sudden increase in force, which was a straight line until the specimen broke. The destruction of the joint occurred rapidly or gradually, with successive fragments of CFRP textile breaking off of the steel surface.

Diagrams of the maximum force–overlap length and maximum force–adhesive thickness relationships for the test specimens are shown in Figures 7 and 8, respectively.

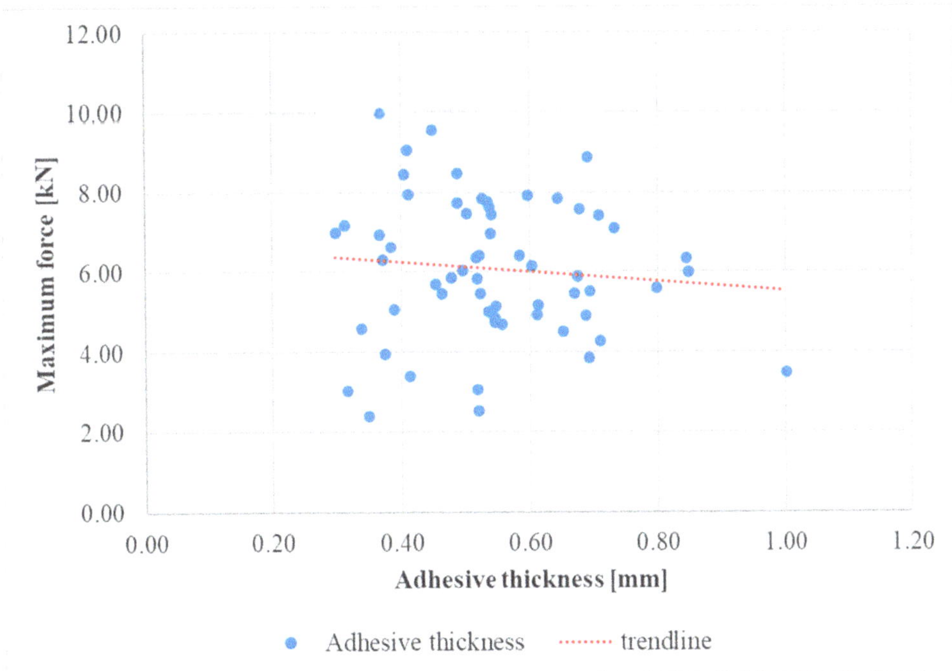

Figure 7. Maximum force-adhesive thickness diagram for all series.

The data points in Figure 7 represent various specimens with differing adhesive thicknesses. The observed trend indicated that greater adhesive thickness resulted in a decreased maximum force capacity. This implies that thicker adhesive layers may compromise joint strength, potentially due to defects within the adhesive.

Meanwhile, Figure 8 shows the relationship between the maximum force applied to the specimens and the overlap length of the bonded joints, and the data points represent different series of specimens with varying overlap lengths. The diagram reveals that as the overlap length increased, the maximum force that the joint could withstand also increased. This indicates that a longer overlap length enhanced the strength of the bonded joint, likely due to the larger bonded area distributing the applied load more effectively and reducing stress concentrations.

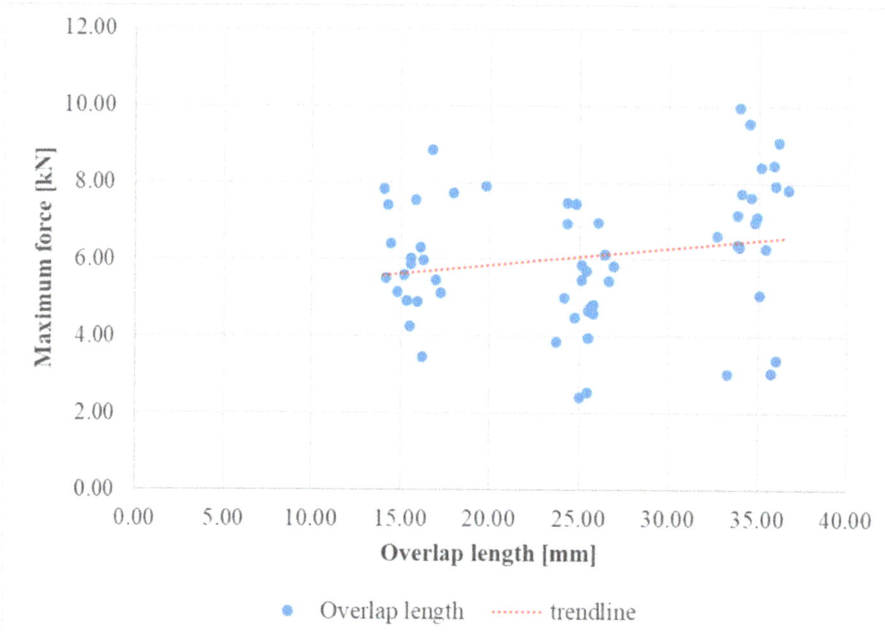

Figure 8. Maximum force-overlap length diagram for all series.

2.3. *Observation of Connection Structure Damage*

Observations of the nature of fracture for CFRP-Steel adhesive connection were conducted. A visual inspection was carried out. Figure 9 shows examples of damaged samples.

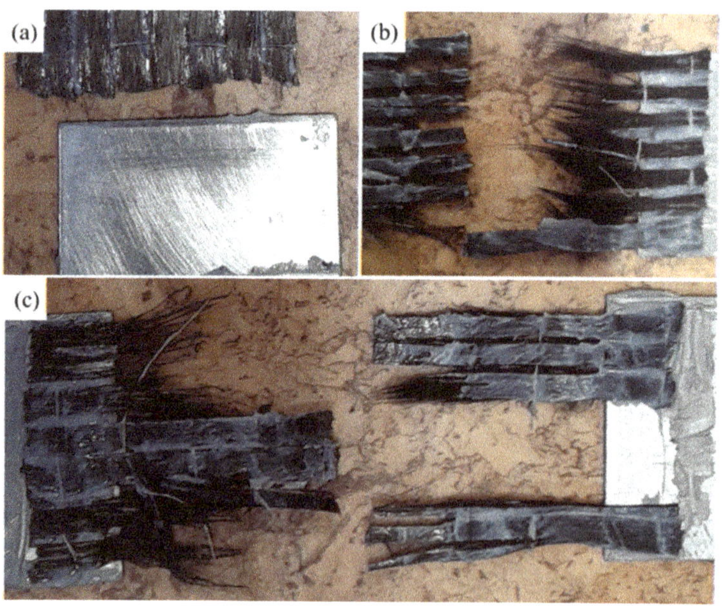

Figure 9. Observed forms of damage: (**a**) adhesive failure at steel/adhesive interface, (**b**) fabric rupture, (**c**) mixed failure.

Upon visual inspection of the damaged samples, different types of damage were observed. For some samples, damage was only observed at the steel/adhesive interface, and for others, fibre rupture. Others were distinguished by mixed damage behaviour. However, in none of the samples, cohesive failure in the adhesive layer or delamination was observed. A full description of the failure mechanisms for each sample is presented in Table 5, where A denotes adhesive failure at the steel/adhesive interface and F corresponds to fabric rupture.

Table 5. Damage mechanisms for each sample.

Sample Label	Damage Mechanism	Sample Label	Damage Mechanism	Sample Label	Damage Mechanism
A1	A, F	B1	A	C1	A, F
A2	F	B2	A	C2	A, F
A3	A	B3	A	C3	F
A4	A, F	B4	A	C4	A, F
A5	A, F	B5	A, F	C5	A
A6	A, F	B6	F	C6	F
A7	A, F	B7	A, F	C7	A, F
A8	A, F	B8	A	C8	A, F
A9	A, F	B9	A	C9	F
A10	A	B10	A, F	C10	F
A11	F	B11	A	C11	F
A12	A, F	B12	F	C12	A, F
A13	A	B13	A, F	C13	A, F
A14	A	B14	A	C14	A, F
A15	F	B15	A	C15	A
A16	A, F	B16	F	C16	A, F
A17	F	B17	A	C17	A
A18	A	B18	A	C18	F
A19	F	B19	A	C19	F
A20	F	B20	A	C20	F

Table 6 presents the average strength of the joint and its standard deviation for each type of failure in different test series.

Table 6. Average joint strength for each type of failure.

Failure	Series A		Series B		Series C	
	Average Strength [kN]	Standard Deviation [kN]	Average Strength [kN]	Standard Deviation [kN]	Average Strength [kN]	Standard Deviation [kN]
A	4.96	0.41	4.44	0.99	3.17	0.16
F	6.32	1.57	6.75	0.67	6.64	0.79
A, F	6.64	1.13	6.60	0.64	8.39	0.79

Figure 10 presents the average force–displacement diagrams for each series.

It should be noted that the joint stiffness and joint strength increased proportionally to the overlap length. Moreover, an increase in the adhesive thickness caused a decreased joint strength. It was also observed that the adhesive failure occurred at the steel interface, indicating that the bond between the steel and adhesive was a critical point. Simultaneously, there was a combination of adhesive failure and fabric rupture, suggesting complex interactions between the materials.

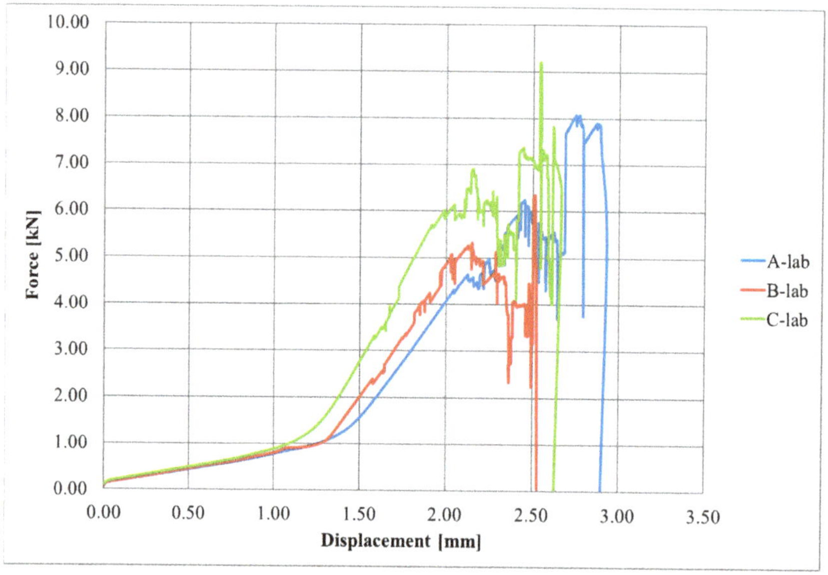

Figure 10. Average force-displacement diagrams for each series—lab.

3. Numerical Analysis

A numerical analysis was conducted in the Abaqus program and later verified and validated based on laboratory test results. Two models were developed, referring to damage mechanisms (Table 5). In the first stage, the goal was to define the material parameters of carbon fibre textile. Second, it was to define the cohesive zone model (CZM) for adhesive connection. The relevant computer hardware and the system type used are presented in Table 7.

Table 7. Computer hardware and system type.

Hardware	Type
CPU	12th Gen Intel(R) Core(TM) i5-12500H 2.50 GHz
RAM	64 GB
System	64-bit operating system, x64-based processor

Figures 11 and 12 detail the steps for setting up a Finite Element Analysis (FEA) simulation for steel and textile adhesive connection. The diagrams outline the process from defining material properties to obtaining output results.

3.1. Numerical Model—Fabric Rupture

ABAQUS [20], a commercial finite element program, was used to simulate the fabric rupture mechanism. The elastic material properties of the carbon fibre textile were taken from [21] with elastic properties given in Table 8. The material behaviour of the textile was defined as elastic–lamina, and the mixed-mode Hashin damage model [22] was implemented with bi-linear behaviour. That model is based on criteria that consider four different damage initiation mechanisms in the form of fibre tension, fibre compression, matrix tension, and matrix compression. The general concept of damage in unidirectional lamina is characterised by the degradation of material stiffness. It holds significant importance in understanding fibre-reinforced composite mechanics. Numerous such materials demonstrate elastic–brittle characteristics; in particular, the initiation of damage in these materials occurs without considerable plastic deformation. Therefore, the influence of plasticity can be dismissed when assessing the behaviour of these materials. It is assumed that the fibres within the fibre-reinforced composite are arranged in a parallel orientation.

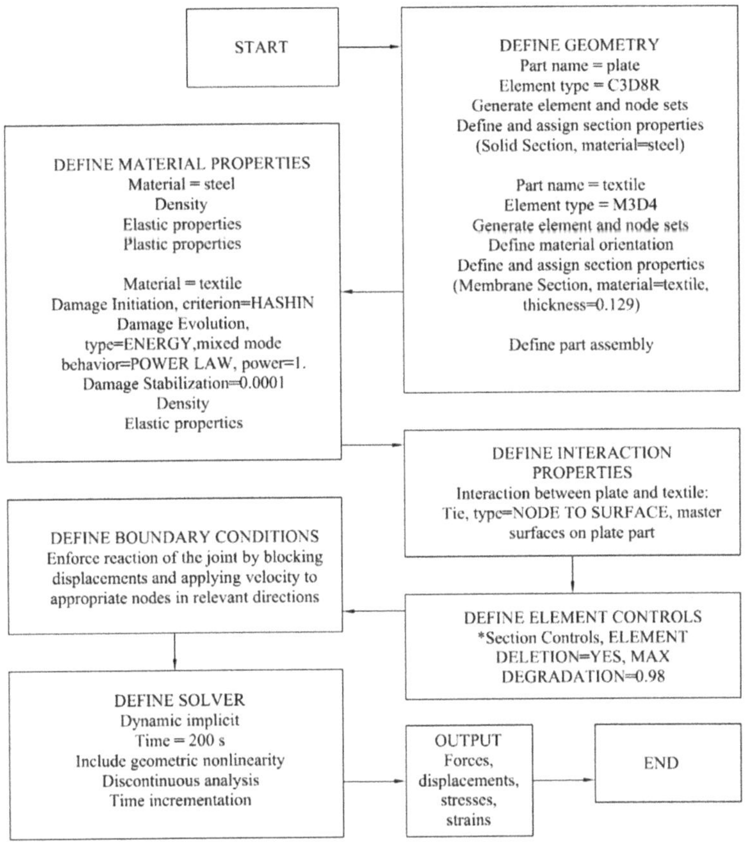

Figure 11. Block diagram for numerical model of textile.

Based on the current research reports, four distinct failure mechanisms appear: tensile fibre rupture, compressive fibre buckling and kinking, transverse tensile and shear-induced matrix cracking, and transverse compressive and shear-induced matrix crushing. The analysis in ABAQUS utilised the failure initiation criteria proposed by Hashin and Rotem [23] as well as Hashin [24], which express the failure surface in the effective stress space. In this model, longitudinal strength was taken from manufacturers data [18], and the remaining parameters were identified during the verification and validation of the model based on the laboratory tests.

After the homogenisation of the fabric, it was assumed that the fibres were contained within the matrix as defined by the Hashin model, with the material properties given in Table 9. However, it should be pointed out that in this analysis, the matrix was modelled as air (Figure 13).

Given that the matrix was air, the transverse characteristics should be equal to "0"; nevertheless, this will result in dividing by "0", and the load bearing capacity of the fabric would be equal to infinity. For this reason, these characteristics must be as small as possible; however, it is necessary to validate the model based on the laboratory results. The value of the parameters was defined as 5 MPa, which was equal to 0.16% of the longitudinal strength of 3200 MPa.

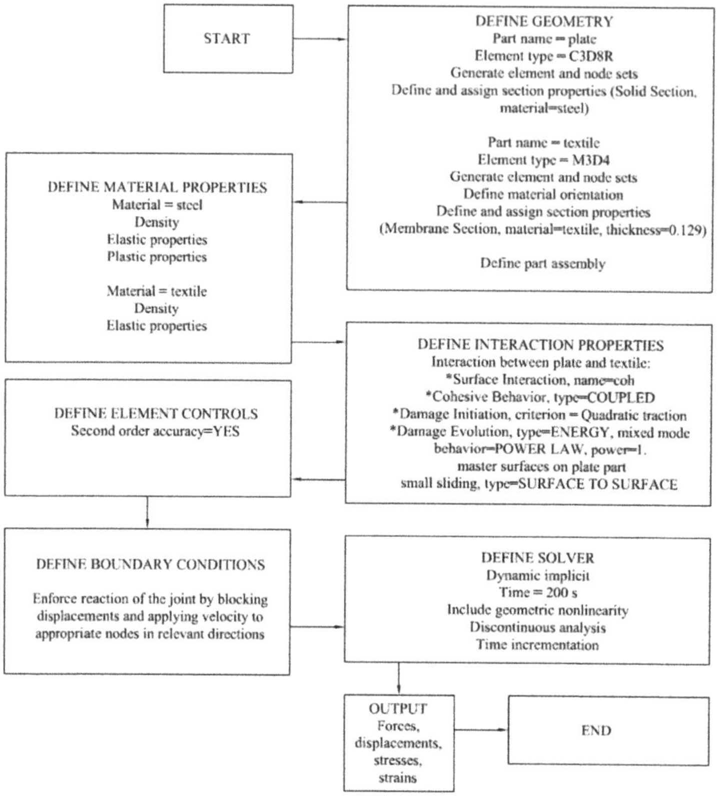

Figure 12. Block diagram for numerical model of adhesive.

Table 8. Carbon fibre textile—elastic material properties.

Parameter	Quantity	Unit
Density	1.83×10^{-6}	kg/mm^3
Elastic modulus of fabric E1	220,000	MPa
Elastic modulus of fabric E2	15,750	MPa
Longitudinal and transverse Poisson's ratio	0.3	[-]
Shear modulus G12	8730	MPa
Shear modulus G13	11,650	MPa
Shear modulus G23	5615	MPa

Table 9. Carbon fibre textile—Hashin damage material properties.

Parameter	Quantity	Unit
Longitudinal Tensile Strength	3200	MPa
Longitudinal Compressive Strength	3200	MPa
Transverse Tensile Strength	5	MPa
Transverse Compressive Strength	5	MPa
Longitudinal Shear Strength	5	MPa
Transverse Shear Strength	5	MPa
Fracture Energy	0.01	N/mm

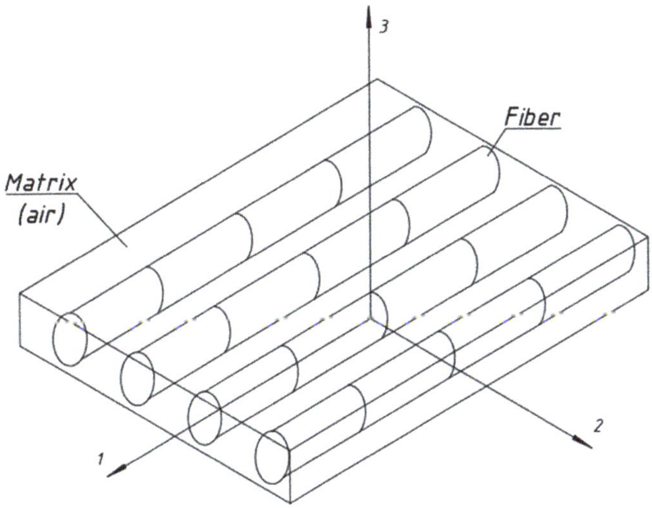

Figure 13. The ABAQUS anisotropic damage model, which is based on research conducted by Matzenmiller et al. [24], Hashin and Rotem [22], Hashin [23], and Camanho and Davila [25].

The calculations were conducted using the dynamic implicit procedure, accounting for geometrical nonlinearity and a time period of 200 s. The boundary conditions applied are presented in Figure 14. On the bottom edge, the displacements were all blocked. Similarly, on the top edge, the displacements were blocked, except the vertical displacement. The displacement velocity was set to 0.01 mm/s in order to enforce tension. Tie connection was used between the plate and textile.

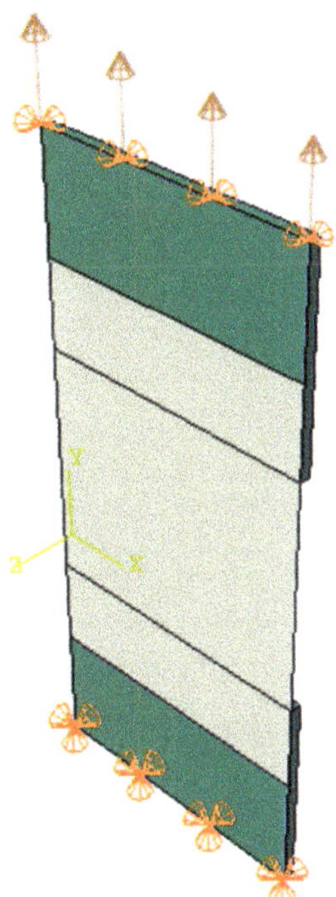

Figure 14. Applied boundary conditions.

An extended mesh sensitivity analysis was performed. Moreover, the impact of the desired material properties (considering the Hashin damage definition) was included. The individual cases are summarised in Table 10.

Table 10. Mesh sensitivity analysis.

Case	Textile Mesh	Plate Mesh	Number of Elements	Number of Nodes	Hashin Damage Parameters [MPa]	Total CPU Time [s]
I	Linear quadrilateral M3D4	Linear hexahedral C3D8R	70,000	85,951	3	2.92×10^5
II	Linear quadrilateral M3D4	Linear hexahedral C3D8R	10,500	14,637	3	9.49×10^3
III	Linear quadrilateral M3D4	Linear hexahedral C3D8R	70,000	85,951	5	2.75×10^5
IV	Linear quadrilateral M3D4	Linear hexahedral C3D8R	112,000	128,191	5	4.04×10^5
V	Quadratic quadrilateral M3D8R	Linear hexahedral C3D8R	10,500	21,757	5	1.45×10^4
VI	Quadratic quadrilateral M3D8R	Linear hexahedral C3D8R	21,000	53,497	5	4.17×10^4

3.2. Numerical Model—Adhesive Connection

In engineering practices, it is assumed that the destruction of the connection should not be of an adhesive nature. Therefore, in order to prevent adhesive destruction, it is necessary to properly prepare the surfaces of the joined elements. On the other hand, the commonly known cohesive zone model (CZM), as the name suggests, allows for the simulation of cohesive destruction, i.e., in which the crack develops in the adhesive layer. Based on the CZM, the authors of this paper attempted to determine a model that would correspond to adhesive destruction at the adhesive–steel interface, which was observed in the laboratory. The ABAQUS program was used to implement a coupled CZM. Given the extremely thin adhesive layer, local effects within the layer were disregarded. A surface-based cohesive behaviour was utilised, characterised by a mixed-mode cohesive law, using the power law criterion. By constructing traction–separation behaviour with coupled and specified stiffness coefficients, the cohesive behaviour was defined. The quadratic traction criterion was used to define the damage initiation criterion, and an energy type with linear softening behaviour was identified as the damage evolution behaviour. We can characterise the initial stiffness matrix as follows:

$$\boldsymbol{K} = \begin{bmatrix} K_{nn} & 0 & 0 \\ 0 & K_{ss} & 0 \\ 0 & 0 & K_{tt} \end{bmatrix} = \begin{bmatrix} 90 & 0 & 0 \\ 0 & 35 & 0 \\ 0 & 0 & 35 \end{bmatrix} [\text{MPa/mm}] \quad (2)$$

where K_{nn}, K_{ss}, and K_{tt} are the elastic stiffnesses in the normal and the two-shear directions, respectively. K_{nn} is equal to the initial slope of the traction–separation model for mode I and can be described as follows:

$$K_{nn} = \alpha \frac{E_a}{T_a} \quad (3)$$

where:

α—a parameter proposed by the authors to bring the destruction mechanism closer to that observed in the laboratory, $\alpha = 0.01$ $[-]$;

E_a—tensile elastic modulus of an adhesive;

T_a—thickness of the adhesive.

Both K_{ss} and K_{tt} are the same and are equal to the initial slope of the traction–separation model for mode II loading and can be described as follows:

$$K_{ss} = K_{tt} = \alpha \frac{G_a}{T_a} \quad (4)$$

where:

α—a parameter proposed by the authors to bring the destruction mechanism closer to that observed in the laboratory, $\alpha = 0.01$ [−]; G_a—shear modulus of an adhesive;

T_a—thickness of the adhesive.

It was assumed that surface preparation, which affects adhesive failure, has the same impact on the elastic stiffnesses in all directions. For this reason, the correction factor in formulae 3 and 4 is the same. The stiffness reduction was intended to simulate the uniform distribution of stresses along the length of the adhesive and, consequently, the sudden development of joint failure. The damage properties of cohesive surface are presented in Table 11. As mentioned, the CZM refers to the failure inside the adhesive. To simulate the adhesive failure, a reduction parameter of $\beta = 0.2$ [−] was proposed. The maximum nominal stress in all directions was defined as follows:

$$t_n = t_s = t_t = \beta R \quad (5)$$

where:

t_n, t_s, t_t—normal and shear tractions in each direction;

R—the tensile strength of the adhesive given by the manufacturer, R = 30 MPa.

The symbol in square brackets [−] indicates that the parameters α and β are dimensionless.

Table 11. Cohesive surface properties implemented in FEM model.

Parameter	Quantity	Unit
Maximum nominal stress in all directions	6	MPa
Normal fracture energy	0.01	N/mm
First and second shear fracture energy	3.8	N/mm
Power law coefficient	1	-

The calculations were conducted using a dynamic implicit procedure accounting for geometrical nonlinearity and a time period of 200 s. The boundary conditions and material properties are presented in Section 3.1. In this case, fabric rupture was not observed; hence, the Hashin damage was not defined.

For the carbon fibre textile, the linear quadrilateral membrane elements of type M3D4 was used, and for steel plates, the linear hexahedral elements of type C3D8R was used. The size of the FE mesh set to 0.4 mm consisted of 131,875 elements and 156,744 nodes.

3.3. Verification and Validation of Numerical Models

In order to verify the effectiveness and quality of the proposed model, validation and verification were carried out (Figure 15). Unfortunately, it was found that the fabric model was very sensitive to the finite element mesh density. With smaller mesh densities, the initiation of destruction occurred faster. It should be emphasised that due to computational limitations, it was not possible to determine a mesh small enough not to affect the calculation result. The smallest tested finite element had a dimension of 0.25 mm.

However, in the case of the model for an adhesive connection, the mesh density did not have a significant impact on the obtained results. In common practice, there should be at least five integration points in the thickness of the solid element, and for this reason, the mesh size was set to 0.4 mm. For a smaller mesh value (0.3 mm), the force–displacement graph coincided with the base graph (0.3 mm). Therefore, it can be concluded that the presented model was not sensitive to grid density, which is recommended in engineering practice. The total CPU time for a model with a 0.4 mm mesh size was equal to 4.18×10^3 s. The value of the maximum force was close to that obtained from the laboratory (Figure 16).

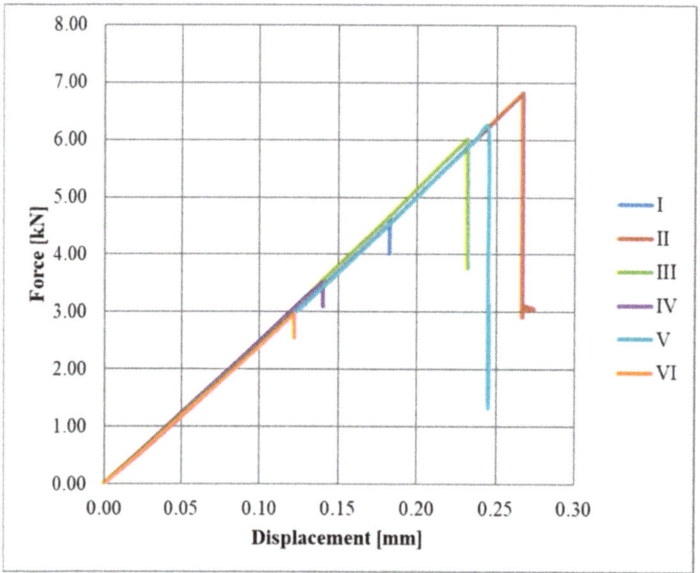

Figure 15. Verification and validation of the fabric rupture model.

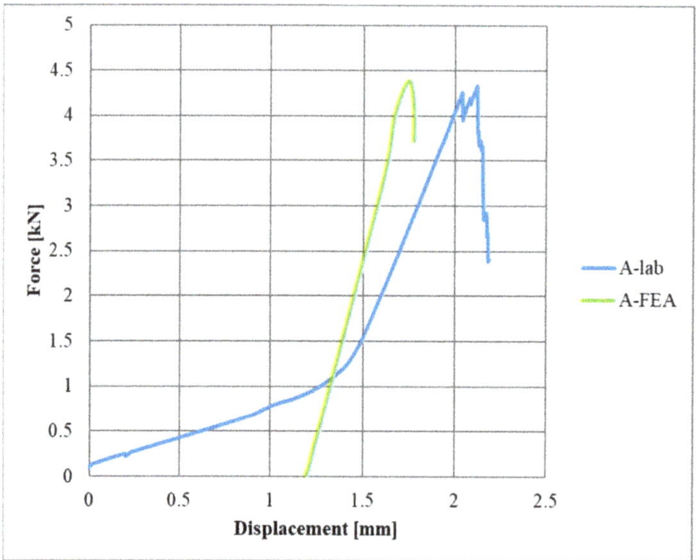

Figure 16. Verification and validation of CZM.

4. Conclusions

In this paper, the laboratory tests and numerical analysis of a composite bonded joint consisting of galvanised steel elements, an adhesive layer, and CFRP fabric were presented. Different overlap lengths (15 mm, 25 mm, and 35 mm) were taken into account in a shear test conducted on 2 mm thick hot-dip galvanised steel plates of S350 GD overlapped on one side with SikaWrap 230 C carbon fibre textile (CFT) using SikaDur 330 adhesive. Based on the conducted analysis, three forms of joint damage were observed, namely, at the steel–adhesive interface, fibre rupture, and mixed damage behaviour. There was a combination of adhesive failure and fabric rupture, suggesting complex interactions between the materials. It was also noted that most danger adhesive failure occurred at the steel interface at the bond between the steel and adhesive layer. Moreover, it was found that the joint stiffness and strength increased proportionally to the overlap length. A simultaneous increase in adhesive thickness caused a decrease in joint strength. As far the numerical composite model consisting of adhesive and fabric models, unfortunately, the fabric model was very sensitive to the density of the finite element mesh. The smaller the mesh size, the greater the fabric destruction parameters that must be applied. With smaller mesh sizes, the onset of destruction occurred faster. But it should be emphasized that the use of the authors' own correlation coefficients α and β allowed for the correct mapping of adhesive damage in the numerical model.

Author Contributions: Conceptualization, M.A.D., K.R. and Z.L.; methodology, M.A.D. and Z.L.; validation, M.A.D.; formal analysis, M.A.D. and K.R.; investigation, M.A.D. and S.A.; resources, K.R. and Z.L.; data curation, M.A.D.; writing—original draft preparation, M.A.D.; writing—review and editing, K.R.; visualization, M.A.D.; supervision, K.R. All authors have read and agreed to the published version of the manuscript.

Funding: This paper was financially supported by Poznan University of Technology: 0412/SBAD/0080, 0412/SBAD/0081.

Institutional Review Board Statement: Not applicable.

Informed Consent Statement: Not applicable.

Data Availability Statement: The original contributions presented in the study are included in the article, further inquiries can be directed to the corresponding author.

Conflicts of Interest: The authors declare no conflicts of interest.

References

1. Dillard, D.A. (Ed.) *Advances in Structural Adhesive Bonding*; Elsevier: Amsterdam, The Netherlands, 2023.
2. Autumn, K.; Sitti, M.; Liang, Y.A.; Peattie, A.M.; Hansen, W.R.; Sponberg, S.; Kenny, T.W.; Fearing, R.; Israelachvili, J.N.; Full, R.J. Evidence for van der Waals adhesion in gecko setae. *Proc. Natl. Acad. Sci. USA* **2002**, *99*, 12252–12256. [CrossRef] [PubMed]
3. Richhariya, V.; Tripathy, A.; Carvalho, O.; Nine, J.; Losic, D.; Silva, F. Unravelling the physics and mechanisms behind slips and falls on icy surfaces: A comprehensive review and nature-inspired solutions. *Mater. Des.* **2023**, *234*, 112335. [CrossRef]
4. Bogue, R. Biomimetic adhesives: A review of recent developments. *Assem. Autom.* **2008**, *28*, 282–288. [CrossRef]
5. Benilov, E. Does the van der Waals force play a part in evaporation? *Phys. Fluids* **2023**, *36*, 032105. [CrossRef]
6. Koguchi, H. Adhesion Analysis considering van der Waals force. *Nenji Taikai* **2022**. [CrossRef]
7. Hajizadeh, K.; Mehdian, H.; Hajisharifi, K.; Robert, E. A van der Waals force-based adhesion study of stem cells exposed to cold atmospheric plasma jets. *Dent. Sci. Rep.* **2022**, *12*, 12069. [CrossRef] [PubMed]
8. Narayanan, A.; Dhinojwala, A.; Joy, A. Design principles for creating synthetic underwater adhesives. *Chem. Soc. Rev.* **2021**, *50*, 13321–13345. [CrossRef] [PubMed]
9. Yamaguchi, T.; Akamine, A.; Sawae, Y. On/off switching of adhesion in gecko-inspired adhesives. *Biosurf. Biotribol.* **2021**, *7*, 83–89. [CrossRef]
10. Baik, S.; Lee, H.J.; Kim, D.W.; Min, H.; Pang, C. Capillarity-Enhanced Organ-Attachable Adhesive with Highly Drainable Wrinkled Octopus-Inspired Architectures. *ACS Appl. Mater. Interfaces* **2019**, *11*, 25674–25681. [CrossRef] [PubMed]
11. Jamali, M.; Tafreshi, H.V. Studying droplet adhesion to fibers using the magnetic field: A review paper. *Exp. Fluids* **2021**, *62*, 161. [CrossRef]
12. Hamasaki, H.; Hirahara, K. The van der Waals cohesive force between two carbon nanotubes. *Appl. Phys. Express* **2023**, *16*, 035002. [CrossRef]
13. Klauser, W.; Bartenwerfer, M.; Fatikow, S. Measurement of sub-nanonewton forces inside a scanning electron microscope. *Rev. Sci. Instrum.* **2020**, *91*, 043701. [CrossRef] [PubMed]
14. Versaci, M.; Angiulli, G.; La Foresta, F.; Crucitti, P.; Laganí, F.; Pellicanò, D.; Palumbo, A. Innovative Soft Computing Techniques for the Evaluation of the Mechanical Stress State of Steel Plates. In Proceedings of the International Conference on Applied Intelligence and Informatics, Reggio Calabria, Italy, 1–3 September 2022; pp. 14–28. [CrossRef]

15. da Silva, L.F.M.; Campilho, R.D.S.G. *Advances in Numerical Modeling of Adhesive Joints*, 1st ed.; Springer: Berlin/Heidelberg, Germany, 2011. [CrossRef]
16. Khan, M.A.; Tipireddy, R.; Dattaguru, B.; Kumar, S. Stochastic modeling of functionally graded double lap adhesive joints. *Mech. Mater.* **2023**, *177*, 104553. [CrossRef]
17. Ramalho, L.D.C.; Sánchez-Arce, I.J.; Gonçalves, D.C.; Belinha, J.; Campilho, R.D.S.G. Numerical analysis of the dynamic behaviour of adhesive joints: A review. *Int. J. Adhes. Adhes.* **2022**, *118*, 103219. [CrossRef]
18. Gcc.sika.com. 2022. Available online: https://gcc.sika.com/content/dam/dms/gcc/j/sikawrap_-230_c.pdf (accessed on 3 September 2022).
19. Usa.sika.com. 2022. Available online: https://usa.sika.com/content/dam/dms/us01/0/sikadur_-330.pdf (accessed on 3 September 2022).
20. Abaqus Unified FEA. 3ds.com. Available online: https://www.3ds.com/products-services/simulia/products/abaqus/ (accessed on 22 November 2022).
21. Dybizbański, M.A.; Rzeszut, K. Experimental and Theoretical Investigation of Galvanized Steel and Fiber-Reinforced Polymer Composites Textile Adhesive Double Lap Joints. *Adv. Sci. Technol. Res. J.* **2023**, *17*, 110–120. [CrossRef] [PubMed]
22. Hashin, Z.; Rotem, A. A Fatigue Criterion for Fiber-Reinforced Materials. *J. Compos. Mater.* **1973**, *7*, 448–464. [CrossRef]
23. Hashin, Z. Failure Criteria for Unidirectional Fiber Composites. *J. Appl. Mech.* **1980**, *47*, 329–334. [CrossRef]
24. Matzenmiller, A.; Lubliner, J.; Taylor, R.L. A constitutive model for anisotropic damage in fiber-composites. *Mech. Mater.* **1995**, *20*, 125–152. [CrossRef]
25. Camanho, P.P.; Davila, C.G. *Mixed-Mode Decohesion Finite Elements for the Simulation of Delamination in Composite Materials*; NASA/TM-2002-211737; NASA: Washington, DC, USA, 2002; pp. 1–37.

MDPI AG
Grosspeteranlage 5
4052 Basel
Switzerland
Tel.: +41 61 683 77 34

Materials Editorial Office
E-mail: materials@mdpi.com
www.mdpi.com/journal/materials

www.ingramcontent.com/pod-product-compliance
Lightning Source LLC
Chambersburg PA
CBHW061555210926
51850CB00009B/2106
9783725875108